DECK CONSTRUCTION

BASED ON THE 2009 INTERNATIONAL RESIDENTIAL CODE®

GLENN G. A. MATHEWSON

Deck Construction Based on the 2009 International Residential Code

ISBN: 978-1-58001-880-7

Cover Art Design:	Dianna Hallmark
Project Manager:	Peter Kulczyk
Project Editor:	Jodi Tahsler
Publications Manager:	Mary Lou Luif
Typesetting/Interior Design:	Sue Brockman

Errata on various ICC publications may be available at www.iccsafe.org/errata.

Definitions listed as (McGraw-Hill) are taken from *The McGraw-Hill Dictionary of Engineering*, second edition, copyright 2003 by the McGraw-Hill Companies, Inc.

First Printing: September 2009

PRINTED IN THE U.S.A.

Table of Contents

Preface

Deck Construction Based on the 2009 International Residential Code is designed to put provisions of the 2009 IRC in the hands of the deck industry in a customized format that includes explanations and illustrations related specifically to deck construction. This book provides the IRC code sections relevant to deck design and installation, making reference to the IRC sections and use of the IRC sections much easier and more recognizable to the deck industry. Deck contractors, installers, designers, homebuilders, inspectors, plan reviewers and manufacturers will all find this book helpful in completing their work in accordance with the IRC provisions, and in enhancing their comprehension of how the IRC applies to deck design and construction.

This book is not written in the format of a "how-to" book and does not intend to teach novice deck builders about the fundamentals of deck construction. Rather, individuals who have a basic understanding of deck construction can now take their skills to the next level, with a deck-specific comprehension of the leading residential code adopted in the United States. Unlike other deck books on the market, either with design ideas or how-to advice, the information contained in this book is organized in a manner that allows the reader to see and compare the author's explanations and the IRC text simultaneously, thus increasing the credibility of the information herein. This book can be used for training and simply read cover to cover, or also can function as a jobsite reference of the IRC during plan approval, construction and inspections.

Nine chapters in this book take the reader through an understanding of how the codes work, how a deck affects the structure it is built against and the building blocks of a deck, including foundations, framing, decking, stairs and guards. It concludes with a discussion of popular amenities often included in deck construction and a brief overview of the differing regulations of commercial decks in public locations. Relevant IRC code sections and IRC defined terms are provided for each topic, followed by a discussion of how the section applies specifically to deck design and construction. In some of the IRC sections provided, the information specific to the following discussion has been emphasized with an underline. Full-color photos and figures illustrate both correct and incorrect installations and intend to give the reader a clear understanding of installation practices to employ and avoid. Building, plumbing, mechanical and electrical provisions that may relate to a new deck installation are included.

The commentary and opinions set forth in this text are those of the author and do not necessarily represent the official position of the ICC. In addition, they may not represent the views of any enforcing agency, as such agencies have the sole authority to render interpretations of the IRC.

About the International Residential Code

Building officials, design professionals, contractors and others involved in the field of residential building construction recognize the need for a modern, up-to-date residential code addressing the design and installation of building systems through requirements emphasizing performance. The *International Residential Code*® (IRC®), in the 2009 edition, is intended to meet these needs through model code regulations that safeguard the public health and safety in all communities, large and small. The IRC is kept up-to-date through the open code-development process of the International Code Council (ICC).

The ICC, publisher of the IRC, was established in 1994 as a nonprofit organization dedicated to developing, maintaining and supporting a single set of comprehensive and coordinated national model construction codes. Its mission is to provide the highest quality codes, standards, products and services for all concerned with the safety and performance of the built environment.

The IRC is one in a family of *International Codes*® published by the ICC. This comprehensive residential code establishes minimum regulations for residential building systems by means of prescriptive and performance-related provisions. It is founded on broad-based principles that make possible the use of new materials and new building designs. The IRC is a comprehensive code containing provisions for building, energy conservation, mechanical, fuel gas, plumbing and electrical systems. The IRC is available for adoption and use by jurisdictions internationally. Its use within a governmental jurisdiction is intended to be accomplished through adoption by reference, in accordance with proceedings establishing the jurisdiction's laws.

About NADRA®

North American Deck and Railing Association, Inc. (NADRA) is the voice of the deck and railing industry. NADRA is devoted to encouraging safe building practices to the industry and promoting deck safety to the consumer. Members are deck builders, manufacturers, dealers/distributors, wholesalers, retailers and service providers to the deck and railing industry. A not-for-profit trade association, the mission of NADRA is to provide a unified source for the professional development, promotion, growth and sustenance of the Deck and Railing building industry so that members can exceed the expectations of their customers. For more information please visit www.nadra.org.

Acknowledgments

Glenn Mathewson thanks the International Code Council staff members who patiently worked with him and his tenacious opinions and ideas throughout the development of this book, including: Peter Kulczyk, Steve Van Note, John Henry, P.E., and Sue Brockman. He feels that working with this group of professionals over the last year to create this educational book was in many ways educational for him as well.

Glenn graciously thanks the Building Division staff members at the City of Westminster, Colorado, for their positive encouragement and peer reviews during this book's development. Without this group of people by his side every day, sharing their experience and expertise with him, he would never be the code professional he is today.

Glenn extends a tremendous appreciation to Andy Engle, Editor for the *Professional Deck Builder* magazine. Andy has acted as a mentor during Glenn's growth as a technical writer, both before and during the development of this book.

Glenn is also thankful to his wife Jennifer and their three young children for sacrificing time with their husband and daddy for so many evenings that he spent working on this book. Finally, Glenn thanks his father, Christopher Mathewson, for raising him under the expectation that he must "do it right, or don't do it at all," and feels that is the essence of what this book should convey.

About the Author

Glenn Mathewson is currently working as a General Building Inspector for the City of Westminster, Colorado, performing residential and commercial inspections of building, plumbing and mechanical installations, as well as residential electrical inspections. Prior to his current position, he was a deck and remodeling contractor and had the opportunity to design and construct numerous large-scale, custom decks.

Glenn has earned a Master Code Professional Certification through the International Code Council. In addition to his work as a municipal inspector, he is currently writing technical articles for the Professional Deck Builder Magazine and is listed as a contributing editor. Glenn also works as a speaker/trainer, developing and presenting building code-related courses for numerous organizations and events around the country, including a full day course titled, "Building Codes for Building Decks," which was offered at the 2008 Deck Expo in Orlando, Florida.

He attends evening classes at the Metropolitan State College of Denver in pursuit of a B.S. degree in Civil Engineering Technology and expects to graduate in 2013.

Chapter 1: Code Administration

Introduction

Like any set of instructions, the IRC must first be understood before it can be used. The first chapter of the IRC contains the administrative provisions intended to provide an understanding of the code's purpose and how it is to be applied and enforced. Also included are the limitations and authority granted to the building official, the building official's representatives (inspectors, plan reviewers) and the board of appeals.

When the IRC is adopted by state governments or local jurisdictions, the administrative provisions are sometimes rewritten or amended. This chapter of the code is very legal in nature, so it is usually highly scrutinized by the jurisdiction's legal department. With that said, most of the general concepts in this IRC chapter are still adopted.

Part One: Code Authority and Responsibility

Definitions

TOWNHOUSE. (IRC) A single-family *dwelling unit* constructed in a group of three or more attached units in which each unit extends from foundation to roof and with a *yard* or public way on at least two sides.

ACCESSORY STRUCTURE. (IRC) A structure not greater than 3,000 square feet (279 m^2) in floor area, and not over two stories in height, the use of which is customarily accessory to and incidental to that of the dwelling(s) and which is located on the same *lot*.

R101.2 Scope. The provisions of the *International Residential Code for One- and two-family Dwellings* shall apply to the construction, *alteration*, movement, enlargement, replacement, repair, equipment, use and occupancy, location, removal and demolition of detached one- and two-family dwellings and townhouses not more than three stories above *grade plane* in height with a separate means of egress and their *accessory structures*.

Discussion: This code section provides a rather detailed and specific list of activities performed that would be governed by the IRC—basically, everything. The IRC is only applicable to one- and two-family dwellings and townhouses, as all other types of structures are regulated by the IBC and the other trade-specific codes. A means to recognize which buildings are regulated by the IRC, and the IRC provisions contained in this book, is explained in Chapter 2. Accessory structures, such as detached garages, free-standing decks, gazebos or shade structures, are also regulated by the IRC when they are incidental to a primary IRC dwelling. For example, a carpenter's wood shop can be constructed using the IRC provisions only if constructed on a property containing an IRC dwelling.

R101.3 Intent. The purpose of this code is to establish minimum requirements to safeguard the public safety, health and general welfare through affordability, structural strength, means of egress facilities, stability, sanitation, light and ventilation, energy conservation and safety to life and property from fire and other hazards attributed to the built environment and to provide safety to fire fighters and emergency responders during emergency operations.

Discussion: This section is the root of all IRC provisions. It provides a summary of the purpose that all the other IRC sections are intended to fulfill. When the interpretation of a section is difficult or unclear, it must at least be consistent with the purposes defined in this section. Building codes have historically focused on providing life-safety provisions, such as structural strength, fire protection and means of egress, but in recent years have extended their regulatory reach to include energy conservation. To maintain the IRC as the minimum allowable standard, it's important to remember that included in these purposes is affordability. Not everyone can afford the best construction methods and materials available, so the IRC must balance the integrity and safety of the structure with economic affordability. The IRC is a minimum standard.

R102.7.1 Additions, alterations or repairs. *Additions, alterations* or repairs to any structure shall conform to the requirements for a new structure without requiring the existing structure to comply with all of the requirements of this code, unless otherwise stated. *Additions, alterations* or repairs shall not cause an existing structure to become unsafe or adversely affect the performance of the building.

Discussion: In remodels or additions to existing building, only the new or modified structures are required to comply with the provisions of the current code. Although, it's also made clear that the new construction cannot cause the existing structure to become "unsafe," which would be defined as anything noncompliant with the code. Deck additions will likely affect the performance of the existing structure if provisions are not considered at the design stage; Chapter 2 is completely dedicated to those situations. For example, when building a deck near an existing window, the window is not required to be brought up to current energy code standards, but it may need to be safety glazed due to new elements of the deck that are considered "hazardous locations," such as stairways (see Chapter 2).

R102.1 General. Where there is a conflict between a general requirement and a specific requirement, the specific requirement shall be applicable. Where, in any specific case, different sections of this code specify different materials, methods of construction or other requirements, the most restrictive shall govern.

R102.4 Referenced codes and standards. The codes and standards referenced in this code shall be considered part of the requirements of this code to the prescribed extent of each such reference. Where differences occur between provisions of this code and referenced codes and standards, the provisions of this code shall apply.

> **Exception:** Where enforcement of a code provision would violate the conditions of the *listing* of the *equipment* or *appliance,* the conditions of the *listing* and manufacturer's instructions shall apply.

Discussion: All of the construction standards and regulations related to IRC construction (one- and two-family dwellings and townhouses) are not specifically located in the IRC alone. The two sections above describe the hierarchy of regulatory authority between IRC provisions, IRC referenced standards and the manufacturer's installation instructions. Generally, all the requirements from the above listed documents must be satisfied by the installation. However, in some circumstances different IRC sections may contradict each other, or may contradict referenced standards or manufacturer installation instructions. "Referenced standard" is the term used to describe other standards usually published by other organizations that are referenced by the I-Codes and are enforced with the same authority as the adopted code, as they are essentially part of the code.

When requirements from any of the above mentioned documents contradict one another, the most restrictive provision shall apply, although with a few caveats. If in the two contradictory requirements, one is more specific to the exact situation than the other, the more specifically related provision shall apply. This may even occur within different sentences in the same IRC section. Many code sections begin by providing the "general" requirements and will then detail more "specific" requirements for special circumstances. While appearing to contradict, it is only a matter of general versus specific.

The exception provided under Section R102.4 creates another circumstance where the most restrictive may not actually apply. If, in any case, the requirements from the IRC or IRC referenced standards would violate the listing of an appliance or equipment, the manufacturer's installation instructions shall apply. Fortunately, these situations rarely occur in typical deck construction.

An important clarification is provided in Section 102.4 related to the regulatory application of referenced standards. Many standards exist that contain a range of information, yet are only referenced by the IRC in regard to a certain specific topic. The regulatory authority of a referenced standard is limited to the extent in which the IRC has referenced it. Other provisions in a referenced standard may be good practice and worth considering, but are not "part of the IRC" (for example, see the discussion of Section R317.1.1 in Chapter 4).

R106.1.2 Manufacturer's installation instructions.

Manufacturer's installation instructions, as required by this code, shall be available on the job site at the time of inspection.

Discussion: Generally speaking, if a manufactured product has an instruction manual, it must be on-site at the time of inspection. Realistically, if it is going to be used during inspection, it would be wise to have the instructions on-site for use by the installers, and again for the inspector. Too often inspectors are the first people on the jobsite to read the instructions, and a failed inspection is often the unfortunate result.

While this section states "as required by this code," the implications of that are deeper than may appear. The code itself may not require the installation instructions, but if the building official's approval of alternative materials is based on information provided in the manufacturer's published installation instructions, then the installation must comply with those instructions. As discussed later in this chapter, almost every manufactured product in deck construction is an alternative, and therefore must be approved. This would include certain decking materials, hidden fasteners, metal connectors, low-voltage lighting or deck drainage systems, for example.

R201.1 Scope. Unless otherwise expressly stated, the following words and terms shall, for the purposes of this code, have the meanings indicated in this chapter.

R201.2 Interchangeability. Words used in the present tense include the future; words in the masculine gender include the feminine and neuter; the singular number includes the plural and the plural, the singular.

R201.3 Terms defined in other codes. Where terms are not defined in this code such terms shall have meanings ascribed to them as in other code publications of the International Code Council.

R201.4 Terms not defined. Where terms are not defined through the methods authorized by this section, such terms shall have ordinarily accepted meanings such as the context implies.

Discussion: IRC Chapter 2 contains definitions intended to aid the code user in understanding the code-specific use of certain terms. The first four sections of Chapter 2 in the IRC explain how terms used in the IRC are to be defined for complete comprehension of the code provisions. Another hierarchy of authority, similar to codes and referenced standards, is presented amongst these four sections.

When looking to define a term used within the IRC, the first stop is Chapter 2 of the IRC. The IRC definitions are sculpted so that a term is defined in the manner it is intended to be used by the IRC, and this can often be very different from its typical meaning. Additionally, IRC definitions tend to provide greater detail and specifics than normal, basic definitions. Following the Chapter 2 definitions of the IRC, definitions from all other I-Code publications can be used, such as the *International Building Code, Plumbing Code or Mechanical Code.* It is important, however, when using other I-Code defined terms that they are used in the context and construction discipline in which they were defined. For example, definitions for electrical installations may not have the same meaning when used for plumbing installations.

If no I-Code publications contain a definition for a term, the ordinarily defined meaning shall apply. General dictionaries, construction or engineering dictionaries, or definitions from industry standards and organizations may all be used to define a term in these cases.

Related definitions from I-Code publications and general dictionaries are provided at the beginning of each chapter part of this book.

R105.2 Work exempt from permit. *Permits* shall not be required for the following. Exemption from *permit* requirements of this code shall not be deemed to grant authorization for any work to be done in any manner in violation of the provisions of this code or any other laws or ordinances of this *jurisdiction*.

10. Decks not exceeding 200 square feet (18.58 m^2) in area, that are not more than 30 inches (762 mm) above *grade* at any point, are not attached to a *dwelling* and do not serve the exit door required by Section R311.4.

Discussion: The previous versions of the IRC have not exempted any deck construction from requiring a permit; however, the 2009 IRC brings a new exception for decks that fall under specific criteria, allowing them to be constructed without a permit. All of the criteria listed in Item 10, above, must be satisfied for the deck to qualify for the exception. The 30-inch-above-grade (762 mm) criterion is related to the lack of required guards for decks lower than this height. For this reason, the height should be measured in the same manner required in Section R312.1 for establishing when guards are required. (See Chapter 7, Part 2 for details.) If at any point on the deck it exceeds this maximum height, then the entire deck must be permitted. If the deck is not attached to the dwelling and does not serve the egress door required for the dwelling, the effects that the deck construction may have on the existing dwelling are minimized, and thus may not require verification by the local building department. Many "free-standing" decks are provided a separate foundation system to support vertical loads, yet are still connected via a ledger as a means to resist lateral loads. In these instances the deck will not qualify for the exception; it must provide a load path for all possible loading conditions that does not require connection to the existing dwelling in any form. The most difficult criterion to determine for this exception is which door on a dwelling is functioning as the egress door, and verifying that it is not the one accessing the deck in question. The egress door is not necessarily always the front door (see Chapter 2 for details).

In the event that a deck permit is not required, or for whatever reason not obtained, this section still requires all construction to be performed in a manner complying with the code, and compliance can still be enforced by the local building official. Considering that so much of the code varies based on the jurisdiction you may be working in, even though you may not need a permit, you should still contact the building department for clarification of the IRC provisions as adopted or amended by the governing jurisdiction, as well as any other local ordinances.

Part Two: Human Authority and Responsibility

Definitions

APPROVED. (IRC) Acceptable to the *building official.*

R104.1 General. The *building official* is hereby authorized and directed to enforce the provisions of this code. The *building official* shall have the authority to render interpretations of this code and to adopt policies and procedures in order to clarify the application of its provisions. Such interpretations, policies and procedures shall be in conformance with the intent and purpose of this code. Such policies and procedures shall not have the effect of waiving requirements specifically provided for in this code.

Discussion: It's the inevitable truth: two people will read the exact same words and come away with different opinions as to what they mean. Interpretation disagreements occur all the time between all the combinations of owners, contractors, subcontractors, inspectors, plan reviewers, building officials, architects, engineers, fire protection personnel, manufacturers, and just about anyone that has involvement in construction. If codes are involved, and people are reading them, it's just a matter of time before there's a disagreement. The code has provided a solution for this problem: the final authority of interpretation belongs to the local building official. To aid in understanding and applying the interpretations, the building official can implement internal policies and procedures, such as standard and generic dimensions for deck foundations in lieu of a complete soil analysis or load-specific-sized piers.

This section certainly provides a lot of authority, power and responsibility to the building official, but it's not without boundaries. The interpretations, policies and procedures must be legitimate. They have to reflect the true purpose and intent of the code. The purpose is general and is described in Section R101.3, previously discussed. The intent, however, is more specific to each code section and each situation. In any attempt to convince building officials that their interpretation is incorrect, you must look at the intent. If word play or a loophole is discovered in the IRC text that seemingly allows for an otherwise nonconforming installation, it can be overruled as not meeting the intent and purpose. In a complete disagreement with the building official's interpretation, an appeal to the board of appeals can be submitted. Details of the board of appeals are discussed later in this chapter.

R104.9 Approved materials and equipment. Materials, *equipment* and devices *approved* by the *building official* shall be constructed and installed in accordance with such approval.

Discussion: This section does not apply to *all* material, equipment and devices; it only applies to *approved* materials, equipment and devices. Any time an alternative material is used, one that is not prescribed specifically in the code, it must be "approved" by the building official. Many code provisions are described in the code as approved, which also requires the building official's approval. When a building official approves the use of material, equipment and devices, it's typically based on evidence that has been provided for the building official's review. Whether based on a calculated design or on test results, the manner in which the material, equipment or device was designed or installed prior to testing is the basis for the approval and the manner in which it must be installed.

R104.9.1 Used materials and equipment. Used materials, *equipment* and devices shall not be reused unless *approved* by the *building official.*

Discussion: Not a very "green" code provision, but this one limits or at least regulates the reuse of material. If the opportunity arises to bid a job using decking that the homeowner salvaged from another site, or even from his existing deck, you need to seek approval first. Once the decking is removed from a deck, it becomes used material and may be required to meet any standards that may then exist for a new installation...even if it is at the same address.

Installing new decking on an existing deck frame would not force the frame to comply with this particular section; it's a structure, not material. It would, however, be required to comply as an altered structure (see the discussion of Section R102.7.1 at the start of this chapter). If the frame were to be dismantled for reuse, you would then have "used material." Often building department personnel may visit a site to approve used material prior to permit approval and issuance.

While decking and framing reuse might draw the most attention, reuse of appliances and devices such as low-voltage lights, hot tubs, fireplaces and other such amenities could also be regulated under this section. In any of these cases it is best to know the model and manufacturer of the used products. Specifications and/or instruction manuals would also be helpful, if not required, for the approval process. The previous section discussed, Section R104.9, would require the used material or equipment to be installed in accordance with any requirements of the approval.

R104.11 Alternative materials, design and methods of construction and equipment. The provisions of this code are not intended to prevent the installation of any material or to prohibit any design or method of construction not specifically prescribed by this code, provided that any such alternative has been *approved.* An alternative material, design or method of construction shall be *approved* where the *building official* finds that the proposed design is satisfactory and complies with the intent of the provisions of this code, and that the material, method or work offered is, for the purpose intended, at least the equivalent of that prescribed in this code. Compliance with the specific performance-based provisions of the International Codes in lieu of specific requirements of this code shall also be permitted as an alternate.

Discussion: Truth be told, this is where most of the construction in our developed communities, and especially in deck construction, is regulated. The IRC contains "prescriptive" provisions for design, material and methods, and these provisions provide a very basic recipe for residential construction. Just as a single cookbook cannot contain every meal and ingredient available for human consumption, neither can a single code book provide all the creative variations conceivable for human occupancy.

An incredibly important, yet seemingly overlooked, sentence in this code section is that the code does not intend to specifically prohibit anything, as long as the intent and purpose of the code is satisfied (see Section R101.3). Consider the cookbook analogy: if you can come up with a way to cook potatoes that creates a safe and edible dish, it's just as good as the scalloped potato recipe in the cookbook and should be approved to serve to your family. Likewise, if there is another way to provide the same purpose and intent of the prescriptive provisions of the code then it should be acceptable, or more specifically "approved."

This is where the challenge may occur—in the approval. It is up to each and every individual building official of each jurisdiction to decide what will be approved in their jurisdiction. "Designs, materials, and methods" in one way or another covers everything the IRC intends to regulate.

Alternative designs are very common, and for decks they are just about mandatory. As you will discover through reading this book, the IRC does not provide a complete package of structural design provisions that apply to decks, which are essentially floors exposed to the weather. As discussed in detail in Chapter 4, most decks are constructed using post-frame construction (post-and-beam), a method of construction not completely detailed in the IRC. For example, providing lateral resistance in post-frame construction without the use of braced walls is not covered in the IRC and would require an approved alternative design (see the discussion of Section R301.1.2 in Chapter 4). The easiest way to gain this approval is through site-specific design. Sometimes a jurisdiction will require a design by a registered design professional if the design falls outside of the prescriptive provisions in the code. State governments regulate the licensure and professional limitations of design professionals. For structural design, some states require a registered professional engineer while others will allow the seal of a registered architect. Regardless of their licensure, a design professional is always bound by professional obligation, responsibility and ethics to perform only design work within the scope of their expertise.

When a building official reviews a structural design for equivalence to the prescriptive design, it is typically not for scrutinizing the calculations behind the design, although they may be staffed to do so. Primarily, a building official is reviewing the overall application of the design. If the design does not violate any other code requirements, is physically possible, actually reflects the conditions of the project and bears an original seal from a design professional, it will likely be approved. Section R301.1.3 discusses more specifically the parameters for engineered design as an alternative (see Chapter 4). There are also pre-engineered documents available from a number of reputable sources that may also be approved as an alternative design, and they may be more affordable and applicable to basic residential deck construction. These documents are discussed in more detail in Chapter 4 and listed in the appendix of this book.

Alternative materials are abundant throughout deck construction, as the days of simple lumber decks are limited. Be it composite plastic, PVC, vinyl, aluminum, fiberglass, stone or tropical hardwood decking or guards; hidden fastener systems, low-voltage lighting, ceramic- or gold-coated hardware or decorative pre-manufactured columns, they are all considered alternative and are subject to the professional opinion of the building official. Testing is the primary method employed to determine the capacities and properties of materials and products. The results of the tests provide the building official with the evidence required to reflect the alternative's equivalency to the prescriptive provisions in the code (see Section R104.11.1 following this discussion).

Alternative methods are a bit more difficult to provide deck-related examples for, so we will have to be a little creative to understand this one. Consider a guard on a deck. It is there to provide fall protection at the edge of the deck. The structural design and connections of the guard fall under "alternative design," and the substance that makes up the guard may be an "alternative material." So what is an "alternative method"? If equivalent fall protection could be provided in some other manner, without the use of a 36-inch (914 mm) high guard at all, it would be an example of an alternative "method." If you think outside the box, you may discover new "methods" that will satisfy the same intent and purpose provided in the prescriptive methods of the IRC. The performance of the built environment is the primary concern of the IRC, not necessarily how it is achieved.

R104.11.1 Tests. Whenever there is insufficient evidence of compliance with the provisions of this code, or evidence that a material or method does not conform to the requirements of this code, or in order to substantiate claims for alternative materials or methods, the *building official* shall have the authority to require tests as evidence of compliance to be made at no expense to the *jurisdiction*. Test methods shall be as specified in this code or by other recognized test standards. In the absence of recognized and accepted test methods, the *building official* shall approve the testing procedures. Tests shall be performed by an *approved* agency. Reports of such tests shall be retained by the *building official* for the period required for retention of public records.

Discussion: The provisions of the IRC are concerned with the performance and function of the built environment, yet provide limited prescriptive means to fulfill these functions. As explained in the previous discussion, most construction features are considered "alternatives." An alternative material or method must be tested in order to provide evidence that it provides an equivalent function to a prescriptive requirement, or is sufficient to perform in a manner required by the IRC, such as in load resistance. While any agency "approved" by the building official is permitted to perform such tests, the International

Code Council Evaluation Services (ICC-ES) is the most prominent and recognized agency in the United States in regard to testing a product for equivalency and satisfaction of IRC provisions.

ICC-ES provides acceptance criteria documents for various construction elements which specify the procedure and comparative basis for testing IRC equivalency. For example, in Chapter 4 of this book it is explained that galvanized fasteners in preservative-treated materials must have zinc-coated weights in accordance with ASTM A 153. If a product manufacturer felt their product could provide equivalent protection to corrosion as provided by zinc coatings in accordance with the IRC referenced standard ASTM A 153, they could have their product tested as an alternative. The acceptance criteria for these product types, developed and used by ICC-ES, provide a test procedure in which the alternative product is tested against a similar product complying with the ASTM standard. If the alternative product was found to provide an equivalent "performance," and the building official was satisfied with the test procedure and results, it could be approved as an alternative in that jurisdiction. In most cases, an Evaluation Services report (ES-report) with positive results will be approved by most building officials, provided the report is applicable to the installation. In this case, the installation must be in accordance with the installation requirements contained in the report. Decking products are the most common alternative found in deck construction, and the details of ICC-ES tests of these products are provided in Chapter 5.

R105.8 Responsibility. It shall be the duty of every person who performs work for the installation or repair of building, structure, electrical, gas, mechanical or plumbing systems, for which this code is applicable, to comply with this code.

Discussion: The implications of this section should be of great interest to anyone performing any construction work. You are responsible. "I've been doing it this way for 30 years," "it passed inspection," "the homeowner wanted it this way," "I didn't know that was required," or any other similar phrase is irrelevant in the eyes of the IRC. It is the responsibility of the installer to comply with the provisions of the locally adopted building code.

R113.1 Unlawful acts. It shall be unlawful for any person, firm or corporation to erect, construct, alter, extend, repair, move, remove, demolish or occupy any building, structure or *equipment* regulated by this code, or cause same to be done, in conflict with or in violation of any of the provisions of this code.

Discussion: If the previous section didn't make the point, this one will. It's not just the installer's responsibility to comply with the IRC; it is "unlawful" not to. While building codes are not initially enforced by police personnel, they are part of the local law, and must be adhered to with the same respect that would be given to any other law.

Part Three: Board of Appeals

R112.2 Limitations on authority. An application for appeal shall be based on a claim that the true intent of this code or the rules legally adopted thereunder have been incorrectly interpreted, the provisions of this code do not fully apply, or an equally good or better form of construction is proposed. The board shall have no authority to waive requirements of this code.

Discussion: While the building official does have a position of considerable authority in regard to the application of the code, there is a system of checks and balances in place to help separate the power. Most jurisdictions employ a Board of Appeals that will hear testimony in argument of a building official's decision. However, there are specific criteria that must be the basis of the appeal to the board: the intent of the code is not the basis of the building official's interpretation, the code does not fully apply to a specific condition of the site or design or an alternative that has been proven to have sufficient evidence of code equivalency has not been accepted. These are the only arguments that will be heard by the board of appeals. It's important to remember that neither the board nor the building official can waive any requirements of the code. Going to the Board of Appeals should be a last resort, only after all attempts to convince the building official have been exhausted.

Chapter 2: The Existing Structure

Introduction

Other than those found at a park site or nature trail, most decks are constructed against or near an existing structure—like a house. A house, or any other type of building, is a complete system; locations of gas and plumbing vents, window openings, glass, lights and many other features are designed to work together to produce a safe environment for human occupancy. Additions and alterations to this "system" can have dramatically negative effects if not properly designed. A deck must be designed thoughtfully, with the features of the existing structure in mind. It's easy to overlook little features like a window location or a gas vent, which could create serious hazards to the occupants' safety if handled improperly. Before the first line of a deck design is put on paper, the locations of the various elements of the existing structure should be noted and analyzed, so that all clearances and other related code provisions can be incorporated into the initial design. In addition to the systems and components, the location on the property and the type of structure the deck will serve must also be considered. Commercial structures, regulated by the IBC and briefly discussed in Chapter 9, require a more in-depth review of the existing building in regard to the occupancy classification and type of construction (see Chapter 9).

Part One: Getting to Know the Site

Definitions

ACCESSORY STRUCTURE. (IRC) A structure not greater than 3,000 square feet (279 m^2) in floor area, and not over two stories in height, the use of which is customarily accessory to and incidental to that of the dwelling(s) and which is located on the same *lot*.

DWELLING. (IRC) Any building that contains one or two *dwelling units* used, intended, or designed to be built, used, rented, leased, let or hired out to be occupied, or that are occupied for living purposes.

DWELLING UNIT. (IRC) A single unit providing complete independent living facilities for one or more persons, including permanent provisions for living, sleeping, eating, cooking and sanitation.

STRUCTURE. (IRC) That which is built or constructed.

TOWNHOUSE. (IRC) A single-family *dwelling unit* constructed in a group of three or more attached units in which each unit extends from foundation to roof and a *yard* or public way on at least two sides.

R101.2 Scope. The provisions of the *International Residential Code for One- and Two-family Dwellings* shall apply to the construction, *alteration*, movement, enlargement, replacement, repair, equipment, use and occupancy, location, removal and demolition of detached one- and two-family dwellings and townhouses not more than three stories above *grade plane* in height with a separate means of egress and their *accessory structures*.

IBC 101.2 Scope. The provisions of this code shall apply to the construction, *alteration,* movement, enlargement, replacement, repair, equipment, use and occupancy, location, maintenance, removal and demolition of every building or structure or any appurtenances connected or attached to such buildings or structures.

> **Exception:** Detached one- and two-family *dwellings* and multiple single-family *dwellings (townhouses)* not more than three *stories* above *grade plane* in height with a separate *means of egress* and their accessory structures shall comply with the *International Residential Code.*

Discussion: When first arriving at a site for deck construction, it's usually pretty easy to determine if it's a commercial occupancy, such as a business, restaurant, or other public area, or a residential occupancy, where people will be sleeping. The IRC does not contain any provisions for any commercial settings, so the *International Building Code* (IBC) is the obvious reference for commercial construction. In residential occupancies, however, knowing what code to use takes a bit more thought. Hotels and apartments that house large numbers of occupants are regulated by the IBC. The safety of these buildings relies on many passive safety elements, such as fire-resistant partitions and corridors, exit stairways and increased exit widths, along with accessibility requirements. Additionally, active features are used in these buildings, such as emergency lighting, fire sprinklers and central fire alarm systems. Residential construction under the IRC does not include many of these features and is therefore much more limiting in occupant load and dwelling unit proximity, yet the absence of these features also simplifies the analysis of the existing structure.

Within the IBC, there are differing, yet justifiable, requirements for different occupancy classifications, just as has historically been the case for model codes. Due to the different uses and numbers of people, it should come as no surprise that a deck used to seat people at a restaurant will require different considerations than an outdoor deck serving an office building. Some provisions for IBC-regulated decks that differ from IRC decks are discussed in Chapter 9.

The IRC is only applicable to one- and two-family dwellings and townhouses and their accessory structures. One-family homes are easy to identify; they are a single structure designed to house a single family. Two-family dwellings are also easy to determine; they are a single structure containing two separate homes, or dwelling units. These homes or "dwelling units" can be arranged and adjoined in any fashion; side-by-side, or top and bottom. Townhouses are a bit more specific. They are essentially one-family homes with no distance between them—connected either side-by-side, or possibly as a four-unit structure back-to-back. Fire-resistant assemblies are required between the townhomes and must extend vertically from the foundation to the roof deck, without any horizontal offsets—they cannot extend over one another like a two-family home. There are a few other specific aspects that define a townhouse (see definitions), but the most apparent is this vertical separation. Basically, if there are more than two connected dwelling units and any one of them has even part of it on top of another, even over a garage, it's not a townhouse and is not regulated by the IRC. (See Examples 2-1 and 2-2.)

Accessory structures are also regulated by the IRC, yet only if the structure is less than 3,000 square feet (279 m^2) in area. Structures larger in area than this are not considered "accessory" to the primary dwelling, and may require additional considerations by the local building official. In accordance with the IRC definitions provided at the beginning of Part 1, a free-standing deck, not attached or abutting the primary structure, will usually be considered an "accessory structure," and would therefore be regulated under the provisions of this code. The distinction of which residential features, other than the primary dwelling, are considered "structures" or "accessory structures" can be difficult to determine using only the general IRC definitions. Determining whether a feature is a "structure" or simply landscaping, and how the IRC applies to it, is the authority of the local building official through the right of interpretation provided in Section R104.1, discussed in Chapter 1 of this book.

Example 2-1: These five townhouses, regulated by the IRC, are separated vertically by fire-resistance-rated walls located just about in line with the white downspouts shown in the photo, but the certainty of that statement comes only after checking the records at the local building department.

Example 2-2: The presence of more than two dwellings without vertical separations means that work on this deck and stairway would be regulated by the IBC (see Chapter 9). Notice the locations of the three front doors to these apartments.

> **Side Note:** Just because a multiple-family development has a name on a sign such as "Townhomes at the Ranch," does not necessarily mean that they are really "townhouses" constructed using the provisions of the IRC.

Type of Construction

The IBC defines and categorizes construction into nine different types: Type IA, IB, IIA, IIB, IIIA, IIIB, IV (Heavy Timber, HT), VA and VB. These types are defined by their ability to resist fire, and are the basis for many differing requirements in the *International Codes*. The IRC does not categorize buildings based on construction type. All structures constructed using the IRC are considered nonrated, and the code allows the use of any material otherwise allowed in the code (combustible or not). IRC structures may still require fire-resistance-rated assemblies for the purpose of dwelling unit separation and exterior wall and opening protection due to proximity to property lines (see "Fire separation distance" in the following discussion). These examples may extend to specific requirements for deck construction, but would not be due to type of construction.

In IBC structures, all nine types of construction can be employed, and this makes analyzing an existing IBC structure for type of construction much more difficult. Most types of construction, other than Type VB will have some effect on the methods used to construct an exterior deck.

Part Two: Location on Property

Definitions

EXTERIOR WALL. (IBC) A wall, bearing or nonbearing, that is used as an enclosing wall for a building, other than a *fire wall*, and that has a slope of 60 degrees (1.05 rad) or greater with the horizontal plane.

FIRE SEPARATION DISTANCE. (IRC) The distance measured from the building face to one of the following:

1. To the closest interior *lot line*; or
2. To the centerline of a street, an alley or public way; or
3. To an imaginary line between two buildings on the *lot*.

The distance shall be measured at a right angle from the face of the wall.

IGNITION-RESISTANT CONSTRUCTION, CLASS 1. (IWUIC) A schedule of additional requirements for construction in wildland-urban interface areas based on extreme fire hazard.

IGNITION-RESISTANT CONSTRUCTION, CLASS 2. (IWUIC) A schedule of additional requirements for construction in wildland-urban interface areas based on high fire hazard.

IGNITION-RESISTANT CONSTRUCTION, CLASS 3. (IWUIC) A schedule of additional requirements for construction in wildland-urban interface areas based on moderate fire hazard.

NONCOMBUSTIBLE MATERIAL. (IRC) Materials that pass the test procedure for defining noncombustibility of elementary materials set forth in ASTM E 136.

PUBLIC WAY. (IRC) Any street, alley or other parcel of land open to the outside air leading to a public street, which has been deeded, dedicated or otherwise permanently appropriated to the public for public use and that has a clear width and height of not less than 10 feet (3048 mm).

WILDLAND-URBAN INTERFACE AREA. (IWUIC) That geographical area where structures and other human development meets or intermingles with wildland or vegetative fuels.

R302.1 Exterior walls. Construction, projections, openings and penetrations of *exterior walls* of *dwellings* and accessory buildings shall comply with Table R302.1.

Exceptions:

1. Walls, projections, openings or penetrations in walls perpendicular to the line used to determine the *fire separation distance*.
2. Walls of *dwellings* and *accessory structures* located on the same *lot*.
3. Detached tool sheds and storage sheds, playhouses and similar structures exempted from permits are not required to provide wall protection based on location on the lot. Projections beyond the *exterior wall* shall not extend over the *lot line*.

4. Detached garages accessory to a *dwelling* located within 2 feet (610 mm) of a *lot line* are permitted to have roof eave projections not exceeding 4 inches (102 mm).
5. Foundation vents installed in compliance with this code are permitted.

TABLE R302.1
EXTERIOR WALLS

EXTERIOR WALL ELEMENT		MINIMUM FIRE-RESISTANCE RATING	MINIMUM FIRE SEPARATION DISTANCE
Walls	(Fire-resistance rated)	1 hour-tested in accordance with ASTM E 119 or UL 263 with exposure from both sides	<5 feet
	(Not fire-resistance rated)	0 hours	≥ 5 feet
Projections	(Fire-resistance rated)	1 hour on the underside	≥ 2 feet to 5 feet
	(Not fire-resistance rated)	0 hours	5 feet
Openings in walls	Not allowed	N/A	<3 feet
	25% maximum of wall area	0 hours	3 feet
	Unlimited	0 hours	5 feet
Penetrations	All	Comply with Section R317.3	<5 feet
		None required	5 feet

For SI: 1 foot = 304.8 mm.
N/A = Not Applicable.

Discussion: The primary incident in American history that lead to the creation of building construction standards was the Chicago fire of 1871, where more than 17,000 structures were destroyed. Since that time, inhibiting fire spread from structure to structure has been a primary purpose of building codes. The IRC and IBC regulate the proximity of buildings to each other using the method of fire separation distance, as defined at the beginning of this section. The purpose of this section is to maintain a specified separation between exterior walls of buildings that have not been constructed using fire-resistant assemblies. In most cases these buildings will be on separate lots, so this distance is divided between both properties and is measured from each structure to the lot line.

This method of building separation provides equal flexibility of development for all property owners, regardless of their neighbors' plans. In the event that a property line is adjacent to a public way, alley or street, it is safe to assume that the need for flexibility in building development is not necessary on that side of the property line, and all the flexibility is granted to the developable lot, possibly allowing construction without fire resistance all the way to the property line (see Example 2-3). In the IRC, for multiple buildings on a single property, an "imaginary" lot line is used to measure the fire separation distance.

While Table R302.1 only lists "walls" and "projections" and does not specifically address decks, the intent and purpose of limiting fire spread still must be provided. In some jurisdictions a deck may be considered a projection, regardless of whether it's cantilevered from the existing structure or supported on one side by posts. In this scenario, nonfire-resistant decks could be no closer than 5 feet (1524 mm) from an interior property line; decks with 1-hour fire-resistant construction on the underside can be built as close as 2 feet (610 mm). Constructing a deck with a 1-hour fire-resistance rating on the underside cannot be easily accomplished using typical deck construction methods and materials (see Example 2-4). The following sections in this chapter are from the *International Wildland-Urban Interface Code™ (IWUIC™)* and provide some other methods of resisting fire spread. These methods may be accepted as an "alternative" and may make deck construction closer than 5 feet (1524 mm) to the property line more feasible. Some building officials may not enforce fire separation distance on ground-level or near-ground-level decks due to the unlikely possibility for a fire to exist beneath the deck, and the difficulty for fire to ignite and spread on the deck surface, but it is always best to check with your local building official.

The general consideration to take from this section, in regard to deck construction less than 5 feet (1524 mm) from a lot line, is to discuss the design and location with the local building department to discover the manner in which they interpret this section, and the applicability of local zoning ordinances.

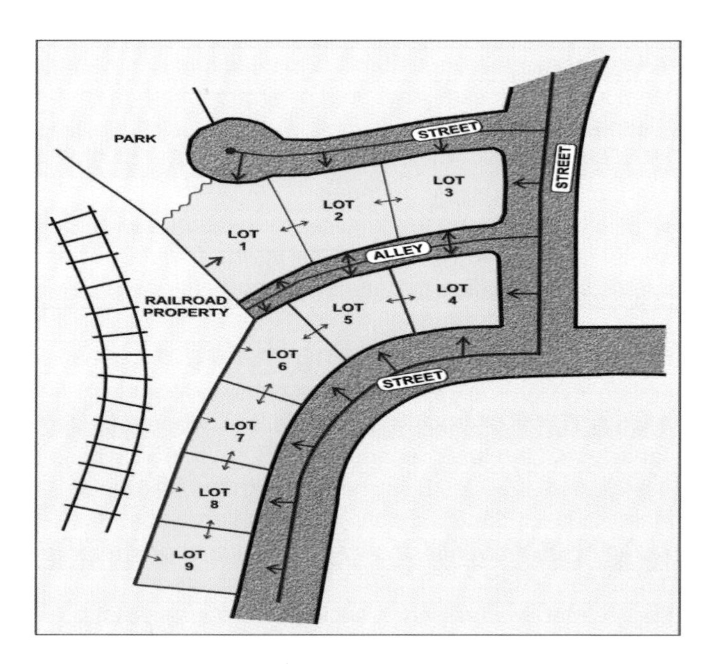

Example 2-3: The red lines in this illustration are the locations that the fire separation distances are measured from, in the direction of the arrows. As specifically stated in the definition, fire separation distance is measured to the closest interior lot line, to the centerline of a street, an alley or public way, or to an imaginary line between two buildings on the lot. When adjacent a public park, such as Lot 1 in this example, the local building department should be consulted for a determination.

Example 2-4: This deck projects 2 feet (610 mm) beyond the side of the house that is located 5 feet (1524 mm) from the property line, and requires 1-hour fire-resistance-rated construction from the underside. This was achieved through the use of 5/8-inch (15.875 mm) type X exterior grade gypsum board attached to the underside of the deck. A water-tight rubber roof membrane was installed on the deck surface and covered with a finished layer of concrete. Typical deck construction cannot generally achieve a fire-resistance rating in this manner.

International Wildland-Urban Interface Code

IWUIC 504.7 and 505.7 Appendages and projections.

Unenclosed accessory structures attached to buildings with habitable spaces and projections, such as decks, shall be a minimum of 1-hour fire-resistance-rated construction, heavy timber construction or constructed of one of the following:

1. *Approved noncombustible* materials,
2. Fire-retardant-treated wood identified for exterior use and meeting the requirements of Section 2303.2 of the *International Building Code*, or
3. Ignition-resistant building materials in accordance with Section 503.2.

503.2 Ignition-resistant building material. Ignition-resistant building materials shall comply with any one of the following:

1. Extended ASTM E 84 testing. Materials that, when tested in accordance with the test procedures set forth in ASTM E 84 or UL 723, for a test period of 30 minutes, comply with the following:
   - 1.1. Flame spread. Material shall exhibit a flame spread index not exceeding 25 and shall show no evidence of progressive combustion following the extended 30 minute test.
   - 1.2. Flame front. Material shall exhibit a flame front that does not progress more than 10 1/2 feet (3200 mm) beyond the centerline of the burner at any time during the extended 30 minute test.
   - 1.3. Weathering. Ignition-resistant building materials shall maintain their performance in accordance with this section under conditions of use. Materials shall meet the performance requirements for weathering (including exposure to temperature, moisture and ultraviolet radiation) contained in the following standards, as applicable to the materials and the conditions of use:
     - 1.3.1. Method A "Test Method for Accelerated Weathering of Fire-Retardant-Treated Wood for Fire Testing" in ASTM D 2898, for fire-retardant-treated wood, wood-plastic composite and plastic lumber materials.
     - 1.3.2. ASTM D 7032 for wood-plastic composite materials.
     - 1.3.3. ASTM D 6662 for plastic lumber materials.
   - 1.4. Identification. All materials shall bear identification showing the fire test results.

Discussion: In an extraordinarily large deck project in the foothills of the Rocky Mountains, a contractor was quite surprised after finding a red-line notation on his approved building permit. The project was in a wildland-urban interface area with increased fire protection requirements. The pressure-preservative-treated 2 by 10 material specified for the joists and beams was not sufficient, and fire-retardant-treated joists and heavy timber beams were the most feasible option. Not fully researching the location of the project and the related expenses before bidding the job put this contractor quite over budget in material costs. This was a mistake I was not soon to forget…

The concern with building a structure in a wildland-urban interface area is the spread of fire from a structure to wildland and from wildland to a structure. While the intent of this code is similar to fire separation distance, the requirements and implications are drastically different. Unlike fire separation distance, this code must be separately adopted, and only regulates certain properties—those in wildland-urban interface areas. There is only one section of this code that directly addresses deck construction. Unfortunately, this single section is not very easy to comply with in deck construction. The provisions are based on the ignition-resistant construction class. The IWUIC provides a method for determining which class of ignition resistance, I, II or III, a deck would have to be constructed under for each property. The ignition-resistant class is generally based on three fundamentals: hazard, defensible space and water supply. The details of these concepts are outside the scope of this text, because for a deck builder, the best way to determine the required class of ignition resistance for a specific site is to ask the local building official. Even decks that are not attached or adjacent to the primary structure may be restricted by the implications of this code.

IWUIC Sections 504.7 and 505.7 are only enforceable when Class I or II ignition-resistant construction is required; Class III has no provisions for deck construction. To limit the movement of fire from nature to buildings, one of the following four fire-resistant methods of construction may be used; each has its own advantages and disadvantages. Some decks may benefit best from a mix of methods. These provisions may also be useful as alternatives to the assemblies listed in Table R302.1 regarding fire separation distance.

One-hour fire-resistance-rated construction: When considering this option, you will find that it is almost always impractical. Fire-resistance ratings are determined in accordance with test procedures detailed in ASTM E 119, by prescriptive designs in IBC Section 720, by calculations in accordance with IBC Section 721, by an engineering analysis comparing to other tested assemblies or by other "approved" methods or designs. For typical residential deck construction, none of these methods are practical—they all use steel, concrete or gypsum products, and would need to be a water-tight assembly. If you have a construction method for a 1-hour fire-resistance-rated deck, you would likely need to seek "approval" from the building official.

Heavy-timber construction: This construction method, also referred to as Type IV, is one of the nine types of construction mentioned in Part 1 of this chapter and detailed further in Chapter 9. Heavy-timber construction provides for excellent fire-resistance. Once the exterior of the timber is charred from the fire, it insulates the core of the timber from the heat, thus increasing its fire-resistant ability. Heavy-timber construction requires minimum nominal widths and depths of solid sawn lumber for three different elements of floor (deck) construction—columns, floor framing and floor decking. Glued-laminated lumber is also allowed, provided its finished width and depth are equivalent to that of solid sawn lumber. Columns, or deck posts, must be a minimum of 8 inches (203 mm) nominal in each direction, and floor framing must be at least 6 inches (152 mm) nominal in width and 10 inches (254 mm) nominal in depth. This is a feasible, yet expensive method for deck construction, until you get to the requirements for the floor surface. Heavy-timber construction requires solid sawn decking at least 3 inches (76 mm) thick, covered by at least 1 inch (25 mm) thick tongue-and-groove flooring, and includes provisions related to moldings fastened to walls. These details are obviously not intended for exterior deck construction and present difficulties in using this method in its entirety for ignition resistance. However, this method may be coupled with ignition-resistant decking as explained further in this discussion.

Approved noncombustible material: It shouldn't be a surprise, but if something is noncombustible, it is not usually a fire danger. Unfortunately, the simplicity of this section is also not much help in deck construction, considering most mainstream decking products and conventional framing methods are of combustible materials. Regardless of the existence of a referenced standard for noncombustibility (see definitions), accepting material as noncombustible may vary from region to region, hence the use of the powerful adjective in this section, "approved." If you believed you are using only noncombustible materials, you would need the specific approval from the local building official.

Fire-retardant-treated lumber (FRT): Depending on material availability, the use of FRT lumber is likely the most familiar method of construction—just a bit more expensive. As in heavy-timber construction, the choice of decking material would be limited to FRT material.

Section 2303.2 of the IBC is referenced for compliant FRT material. Information about this material is provided in Chapter 4

Decking Material: Since typical manufactured decking material does not usually conform to the four previously mentioned methods of construction, the use of Section 503.2, Item 1, (provided prior to this discussion) can be helpful. This section provides various testing procedures that can be used to determine some manufactured decking products' resistance to ignition. Upon verification of a product's testing results, some decking may be approved when Class I or II ignition-resistant construction is required.

Side Note: In the example provided above of the large deck in the wildland-urban interface area, the most feasible and least costly method of compliance came through the approval of a blend of the aforementioned methods. Heavy timber was used for all the posts and beams, FRT lumber for all the joists, and the decking was approved as an alternative to noncombustibility due to the Class A fire rating specified by the manufacturer of the product. It was up to the local code official to approve this method, and luckily...he did!

Government Planning and Zoning Regulations

When considering a deck's location on a piece of property, the building code is not the only requirement that must be addressed. "Setbacks" is a more familiar term to most residential contractors than "fire separation distance." This is most likely because setbacks are usually much greater in distance. Setbacks are generally development or site specific, not jurisdiction specific like the building codes. Different homes in different developments can have different setback distances. Setbacks may also vary from the back, side or front of a house, as well as whether it is a corner lot or lot adjacent a public way, golf course or park. There is no definitive method for determining the setbacks for each lot, other than going straight to the planning or zoning department and asking.

When permitting a deck through the local building department, addressing setbacks is not typically a problem, as the plans must usually be approved by planning/zoning prior to a building department plan approval. Typically, issues only exist when a deck is less than 30 inches (762 mm) above grade and exempt from a required building permit (see Chapter 1). In these instances, a trip to the planning and zoning department is still necessary for constructing a legal deck, as this exception may not exist in the local zoning code. In some cases, if the deck is low enough to not require a building permit, it may not be subject to the zoning setbacks. The only way to know for certain is to contact the local planning and zoning department for every project. While the ICC does publish a model zoning code (*International Zoning Code*), most jurisdictions tend to create their own regulations; and these regulations can vary considerably. In addition to deck location, planning departments may have material type and color requirements for construction visible from public ways. I remember a small ground-level deck project I bid that backed up to a golf course. Bricked deck columns up to guard height were required that would have exceeded the cost of the deck. The deck required no building permit, but the brick did—needless to say, the owners went with a flagstone patio instead.

Side Note: Many planning and zoning ordinances include a maximum number of accessory structures allowed on a single residential property, sometimes allowing only one. If the site already has a detached garage or shed, constructing a detached free-standing deck or gazebo may be prohibited altogether.

Part Three: Landings at Exterior Doors

R311.2 Egress door. At least one egress door shall be provided for each *dwelling* unit. The egress door shall be side-hinged, and shall provide a minimum clear width of 32 inches (813 mm) when measured between the face of the door and the stop, with the door open 90 degrees (1.57 rad). The minimum clear height of the door opening shall not be less than 78 inches (1981 mm) in height measured from the top of the threshold to the bottom of the stop. Other doors shall not be required to comply with these minimum dimensions. Egress doors shall be readily openable from inside the *dwelling* without the use of a key or special knowledge or effort.

Discussion: You may not be aware, but there's a specific exterior door in every home that is different and special from all the other doors. It is the safest door in the house and is the one that you would most likely want to use. It's the "egress door" and it is the only one like it required by the code—all others are just exterior doors, and they're just extra. Of course this door is not labeled or identified, and it's not common knowledge to anyone.

The landing, or deck, on the exterior side of the egress door has a different requirement than landings for all the other exterior doors. When constructing a new deck outside of any exterior door of a home, the deck acts as the landing for the door, thus creating the need to determine which door is the egress door and properly consider its specific requirement into the deck design.

Usually the egress door is the front door, but there is no IRC requirement for this; it could just as well be the back door. If the door leading to the deck is not side-hinged, such as a sliding glass door, then it is not the egress door. Additionally, if it is side-hinged, but is not 3 feet (914 mm) wide and 6 feet, 8 inches (2032 mm) high, it's also not the egress door. With this said, not all existing homes are built to this code, especially because the code is not retroactive, and there may not be any exterior doors that comply with all the provisions for the egress door. In that situation, I would recommend considering the largest side-hinged door as the egress door, or ask the local code official for an interpretation. The following sections discuss the special requirements for the egress door landing as well as requirements for all other exterior door landings.

R311.3 Floors and landings at exterior doors. There shall be a landing or floor on each side of each exterior door. The width of each landing shall not be less than the door served. Every landing shall have a minimum dimension of 36 inches (914 mm) measured in the direction of travel. Exterior landings shall be permitted to have a slope not to exceed $^1/_4$ unit vertical in 12 units horizontal (2-percent).

> **Exception:** Exterior balconies less than 60 square feet (5.6 m^2) and only accessible from a door are permitted to have a landing less than 36 inches (914 mm) measured in the direction of travel.

Discussion: All exterior doors must be provided with a landing on the interior and exterior of the doorway. Just as with all IRC required landings, the door landings must be at least 36 inches (914 mm) in the direction of travel, generally considered the direction at a right angle to the door in the closed position. The landings, both interior and exterior, must only be as wide as the door served. This requirement is not referring to the entire door assembly itself, but rather the actually door opening area. In the case of a sliding glass door with an operable panel adjacent to a fixed panel, the landings must be only at the operable panel, whereas French doors, with two operable panels, would require landings as wide as both door leaves (see Example 2-5). The exterior landings of these doorways are permitted to slope a maximum of a ¼-inch rise for a 12-inch horizontal distance. Similar to landings at stairways and ramps, this allowable slope can inhibit water pooling on non-permeable deck surfaces, or decking installed without gaps.

A new exception to the minimum depth of exterior door landings was included in the 2009 version of the IRC, allowing small balconies, those projecting from the structure without additional support, to be constructed without regard to any minimum depth. However, this exception is only applicable when the balcony is served only by a door from the primary structure and is less than 60 square feet (5.6 m^2) in total area.

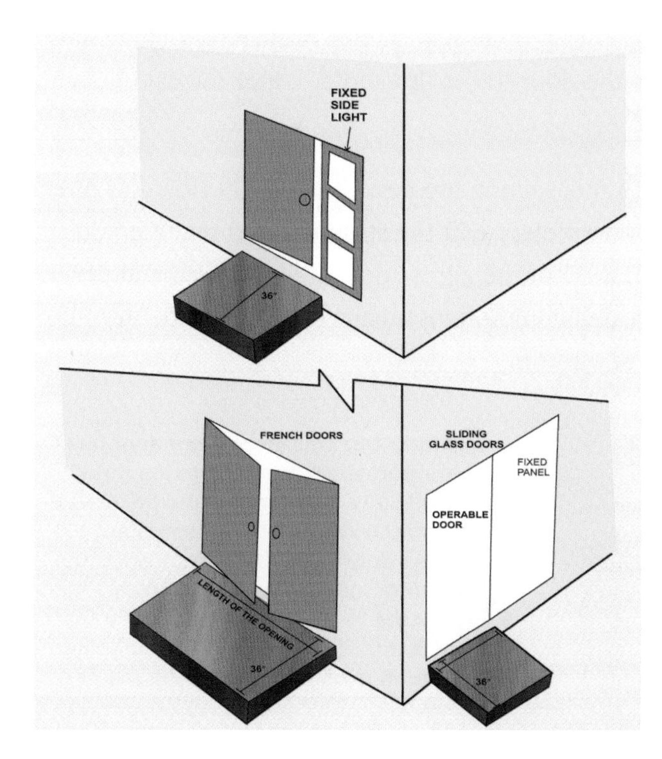

Example 2-5: Only the operable portion of a door assembly is required to be provided a landing. However wide the opening is, a landing of the same width is required.

R311.3.1 Floor elevations at the required egress doors.

Landings or floors at the required egress door shall not be more than 1 1/2 inches (38 mm) lower than the top of the threshold.

Exception: The exterior landing or floor shall not be more than 7 3/4 inches (196 mm) below the top of the threshold provided the door does not swing over the landing or floor.

When exterior landings or floors serving the required egress door are not at *grade*, they shall be provided with access to *grade* by means of a ramp in accordance with Section R311.8 or a stairway in accordance with Section R311.7.

Discussion: The first paragraph in this section is the general requirement for the height of the egress door landings without regard to the direction of the door swing. Without any further consideration, the egress door cannot be more than 1 1/2 inches (38 mm) below the top of the threshold. However, in this section the exception is more commonly the rule. In the majority of homes the primary exterior doors will swing into the dwelling, allowing the exterior landing to be up to 7 3/4 inches (196 mm) below the top of the threshold (see Example 2-6). In regions subject to winter snow accumulation, a step down to the exterior landing is highly desirably. As snow accumulates, the bottom layers compress and become denser. As melting begins, the base of snow accumulation acts like a sponge and holds a considerable amount of liquid water. When these lower few inches of heavily moist snow are against the seal of the door to the threshold, water intrusion is very likely.

The final statements in this section are new in the 2009 IRC and intend to clarify the need to completely exit the structure and reach grade level. The exterior landing of the egress door must either be at grade level, or must be provided a compliant means to reach grade through the use of a stairway or a ramp.

Example 2-6: This front entry door has two risers from the landing up to the threshold, so while it is side hinged and 36 inches (914 mm) wide, it is not a compliant "egress door." The new deck off of the back door may have to be built as a landing for the "egress door" since the front door is not. This would mean one riser maximum from the deck to the top of the threshold.

R311.3.2 Floor elevations for other exterior doors. Doors other than the required egress door shall be provided with landings or floors not more than 7 3/4 inches (196 mm) below the top of the threshold.

> **Exception:** A landing is not required where a stairway of two or fewer risers is located on the exterior side of the door, provided the door does not swing over the stairway.

Discussion: Once the location of the "egress door" has been established, all other exterior doors can be provided an increased flexibility in their landing design. The general requirement for these doors is identical to the exception provided for the "egress door," yet allows the doors to swing over their landings when up to a maximum of 7 3/4 inches (196 mm) below the top of the threshold (see Example 2-7). The exception under this section further increases design flexibility by allowing a stairway with up to two risers to exist on the exterior side of these doors. Under this exception, however, the door cannot swing over the two-riser stairway (see Example 2-8).

Door Landing Elevations

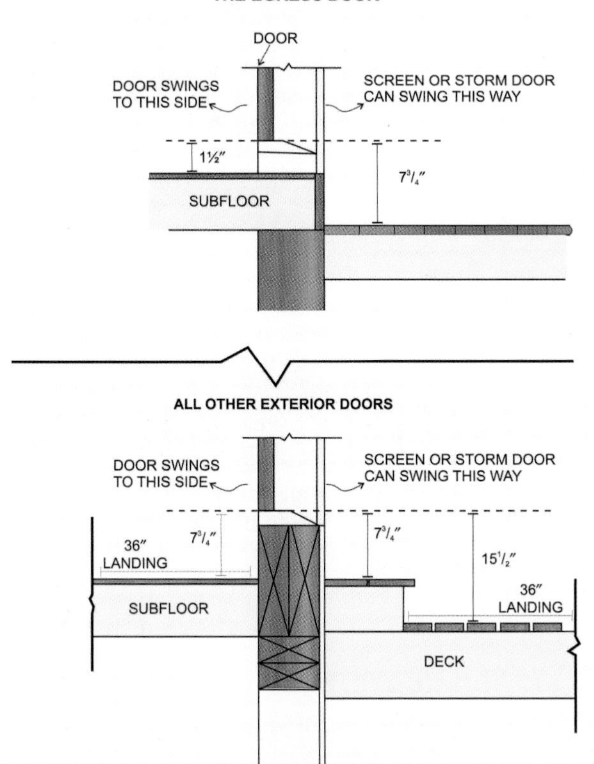

Example 2-7: A large beam, taller than the floor joists, was installed at the exterior rim of this home during construction to carry a concentrated load from the roof. The height of the beam above the floor and its inability to be notched would require the addition of a door at this location to be raised above the floor. Provided this door is not the egress door, it could be installed without the need for a raised landing within the room, provided the dimensions are in accordance with the graphical example provided with the photo.

Example 2-8: Stairs outside of any exterior door cannot exceed two risers from the landing. The four risers in this photo would have to be separated by a 36-inch (914 mm) deep landing at the height of the second tread in order to be compliant with the IRC.

R311.3.3 Storm and screen doors. Storm and screen doors shall be permitted to swing over all exterior stairs and landings.

Discussion: This section is applicable to all exterior doors, including the "egress door." Storm doors and screen doors, which naturally swing towards the exterior, may do so in all cases, regardless of the location of the exterior landing or the presence of steps.

Part Four: Dryer Exhaust Termination

Definitions

Back-draft damper. (McGraw-Hill) A damper with blades actuated by gravity, permitting air to pass through them in one direction only.

M1502.3 Duct termination. Exhaust ducts shall terminate on the outside of the building. Exhaust duct terminations shall be in accordance with the dryer manufacturer's installation instructions. If the manufacturer's instructions do not specify a termination location, the exhaust duct shall terminate not less than 3 feet (914 mm) in any direction from openings into buildings. Exhaust duct terminations shall be equipped with a backdraft damper. Screens shall not be installed at the duct termination.

Discussion: A very common item on the exterior of a house is a clothes dryer exhaust termination hood. These hoods can be easily identified by their backdraft damper (the top hinged cover), or by turning on the dryer and checking for exhausted air. Blocking or obstructing these hoods can cause more problems than just damp towels and long drying time. Dryers that can't exhaust the heat, store the heat…and the lint. With fuel (lint), oxygen and heat, you have the recipe for fire. The US Consumer Product Safety Commission reported 15,600 dryer fires in 1998, with 20 deaths and 370 injuries, not a statistic a deck builder wants to contribute to. In the case of a gas dryer, the products of the gas combustion are also meant to be exhausted via this duct; the high levels of carbon dioxide, and possibly carbon monoxide, present in the fuel-gas exhaust are not usually safe when left inside the building.

The IRC references the dryer manufacturer for installation instructions concerning the exhaust termination requirements. However, a dryer is a relatively temporary appliance, making it somewhat impractical to specifically reference for a more permanent installation, such as a deck. Conveniently, dryer manufacturers' requirements are generally the same. A 12-inch (305 mm) vertical clearance typically is required between the bottom of the exhaust opening and the grade or floor below. Leaves, snow and other debris may collect in front of and beneath the opening, contributing to the collection and concealment of lint and the obstruction of the opening. The location and height of the deck surface, where abutting the home, could adversely affect the code compliance of the existing dryer installation (see Example 2-9).

Dryer exhausts are commonly located at the rim joist of a home's floor; the same place decks are often attached, so it seems to be common practice for dryer exhausts to terminate beneath the deck surface. While this is not regulated by the code or most manufacturers, it would certainly not be a good idea to force warm, moist air beneath a deck if it was close to the ground and had limited ventilation. Even without obstruction, there is a tendency for lint to collect at the backdraft damper mechanism; therefore it's ideal for there to be access to the termination for inspection and maintenance purposes (see Example 2-10). While this is good practice, if the dryer manufacturer does not specify access, it would technically not be required.

If any new openings are being added or altered to the existing structure, such as a new door leading out to the deck, the proximity of dryer exhaust terminations to building openings must be considered. This proximity is discussed in Chapter 8.

Example 2-9: Half of this exhaust termination is already blocked by the deck; a small snow or debris accumulation will take care of the other half.

Example 2-10: Even without foreign obstructions from low grade clearance, lint can create an obstruction and fire hazard. Accessible dryer exhaust terminations are vital to proper building maintenance, yet may or may not be required by the manufacturer.

Part Five: Fuel-Burning Equipment Air Intake and Exhaust

Definitions

COMBUSTION AIR. (IRC) The air provided to fuel-burning *equipment* including air for fuel combustion, draft hood dilution and ventilation of the *equipment* enclosure.

CONDENSING APPLIANCE. (IRC) An *appliance* that condenses water generated by the burning of fuels.

DIRECT-VENT APPLIANCE. (IRC) A fuel-burning *appliance* with a sealed combustion system that draws all air for combustion from the outside atmosphere and discharges all flue gases to the outside atmosphere.

G2407.11 (304.11) Combustion air ducts. *Combustion air* ducts shall comply with all of the following:

2. Ducts shall terminate in an unobstructed space allowing free movement of *combustion air* to the *appliances*.
8. *Combustion air* intake openings located on the exterior of a building shall have the lowest side of such openings located not less than 12 inches (305 mm) vertically from the adjoining grade level.

Discussion: Chapter 24 of the IRC contains provisions for fuel-gas burning appliances, those that burn liquid petroleum gas (LPG or propane) or natural gas, such as furnaces, water heaters and fireplaces. Chapters 12 through 18 of the IRC address similar appliances that burn fuel other than gas, such as wood, pellets or oil. For all these appliances to function correctly and efficiently, they must be provided with a source of fresh air. Combustion air, the topic of the above code section, is referring to this fresh air. Combustion air is required for multiple functions: providing oxygen for the combustion process, makeup air to balance the air being exhausted, dilution air to help produce and maintain a draft in natural draft appliances equipped with draft hoods or draft diverters, such as water heaters, and ventilation air to help cool the room and keep the equipment running efficiently. The chemistry and science behind the need for combustion air is outside of the scope of this book, yet that is not meant to imply that it is of little importance. Often, combustion air is ducted directly to the mechanical room via combustion air ducts that terminate at the exterior of the building. In those instances it is vital to the health and safety of the building occupants that they be considered during any exterior alterations to a building (see Examples 2-11 and 2-12).

Item 8 in the list of requirements must be considered when designing the location of a deck against an existing structure. The requirement for a 12-inch (305 mm) clearance above grade is intended to prevent snow, debris, landscaping and similar exterior elements from obstructing the openings, and is an identical clearance as that required for dryer exhaust termination, as previously discussed. In considering the purpose and intent of this section, it is logical and appropriate to regulate the vertical distance to a walking surface of a deck, patio or any other horizontal surface the same as if it were grade.

Example 2-11: Combustion air intake openings for the "high" and "low" openings in the building must be considered in deck design. The top of a deck against this wall would need to be at least 12 inches (305 mm) below the lowest edge of the openings.

Example 2-12: Combustion air can be obtained easily from underneath decks, provided there are sufficiently sized ventilation openings to that area [see the discussion of Figure G2407.6.1(1) in this part of Chapter 2].

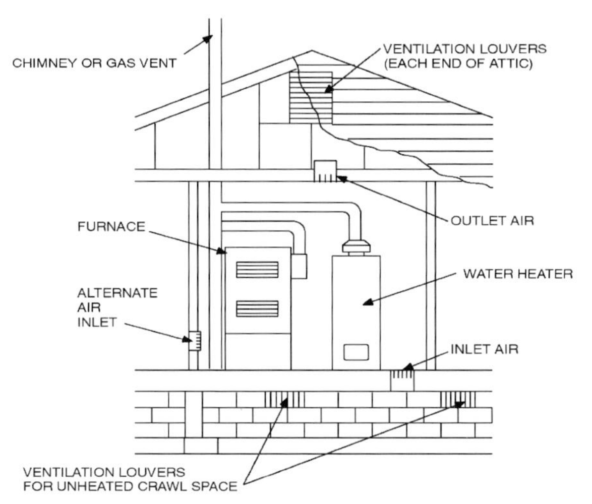

Figure G2407.6.1(1) Combustion air for both gas-burning and liquid- or solid-burning appliances can be obtained from under-floor areas (crawl spaces) provided there are sufficient openings from the under-floor area to the exterior.

Discussion: Generally speaking, the IRC is not referring to under-deck areas in this figure, but according to the intent and purpose of providing sufficient combustion air, there would be little difference. If the combustion air openings are underneath a deck, sufficient airflow to that area must be provided. For a second-story or garden-level deck this likely will not be an issue at all. Ground-level decks, or those skirted with solid material from the framing to the ground, may inhibit airflow under the deck and to the combustion air openings (see Example 2-13). Additionally, the use of tongue-and-groove decking profiles, increasingly present in manufactured decking products, will also inhibit the ventilation traditionally provided by the gaps between the deck boards.

If a deck is to be skirted solid to the ground, or otherwise designed in a manner that limits the under-deck air movement, ventilation openings should be installed in effective locations. If the skirting is provided with gaps between the material or constructed using material with openings, such as lattice, there will likely be sufficient airflow without the use of specific ventilation openings (see Example 2-14). Part 7 of this chapter discusses foundation vents for existing structures and provides IRC sections that should be used for guidance when considering effective locations for these under-deck ventilation openings. Ventilation under the deck will also help reduce moisture collection on the framing and decking materials, and is even required by some decking manufacturers, so when in doubt, err to the side of more ventilation.

Example 2-13: Combustion air openings should not terminate under decks with inadequate ventilation.

Example 2-14: When skirting a deck to the ground, lattice or other materials with free airflow are strongly recommended, not only for combustion air purposes, but also for removing moisture-laden air from beneath the structure.

G2427.6.1 (503.6.1) Installation, general. Gas vents shall be installed in accordance with the terms of their listings and the manufacturer's instructions.

Discussion: While most gas vents penetrate the building through the roof, sometimes vents leave the building through a sidewall. Category IV appliances are high-efficiency appliances with low-temperature flue gases that typically vent through plastic pipe and terminate at exterior walls (see next discussion). In contrast, when common Category I equipment is vented out an exterior wall, it usually must be run up the exterior of the building, where it can extend above the roof surface. In these somewhat rare cases, a metallic vent may need to extend through a deck surface. Depending on the vent type, clearances to any combustible components of the deck construction must be maintained. Type B-vent has an airspace trapped between two walls of the vent, creating an insulating effect and allowing a 1-inch (25 mm) clearance to combustibles (see Example 2-15). Conveniently B-vent is labeled with the clearances right on the vent itself. Single-wall vent, however, is not usually labeled, yet is relatively easy to identify. Single-wall vent, while a rare encounter, would require a 6-inch (152 mm) to 36-inch (914 mm) clearance, depending on the type of equipment being vented. That determination is outside the scope of this publication.

Example 2-15: The cuts around this vent pipe are not completely a result of poor workmanship, but rather compliance with the vent's 1-inch (25 mm) minimum clearance to combustible material. Noncombustible decking material could have been in contact with the vent.

G2425.14 (501.14) Category II, III and IV appliance venting systems.

The design, sizing and installation of vents for Category II, III and IV *appliances* shall be in accordance with the *appliance* manufacturer's installation instructions.

Discussion: Most fuel-burning equipment currently installed in homes across the US is considered Category I. This equipment utilizes a steel vent (B-vent or single wall) and a natural draft or fan-assisted draft to remove the products of combustion (exhaust) from the building, and is discussed in the preceding discussion.

Fuel-gas burning equipment Categories II, III and IV were created to organize other types of fuel-burning equipment based on the exhaust temperature and the possibility for condensation of flue gasses, although currently there is little to no equipment available with the characteristics of Category II and III. Category IV equipment, however, is becoming increasingly popular in new construction as well as equipment replacements in existing residences. Also referred to as "high-efficiency," "90-percenter," "power-venter," or "condensing," Category IV equipment is exactly that, highly energy efficient, with 90 percent or more of the heat energy produced transferred to the conditioned circulating air, vented by a blower and usually causing the exhaust gas to condense inside the vent pipe.

Category IV equipment venting systems are regulated primarily by the appliance manufacturer, due in part to their relatively short history in construction and the inconsistent methods and designs found between manufacturers. Most often PVC pipe vents are used as the exhaust and intake for Category IV equipment. With a fan-powered exhaust system, the need for a generally vertical vent is eliminated, and thus these appliances are usually vented through the rim joist or side wall of a building—right where you would want to put a deck (see Examples 2-16, 2-17 and 2-18). Most manufacturers require similar clearances around vents and intakes, but if available, referencing the appliance's installation instructions would be recommended. Generally there are two configurations that should be investigated when designing a deck near a Category IV appliance exhaust termination. If the deck is below the termination, there is a vertical clearance that must be maintained. This clearance is often 12 inches (305 mm) above grade or anticipated snow accumulation. If the deck is above the termination, there may be a different clearance to consider, oftentimes up to 36 inches (914 mm) vertically (see Example 2-19).

Example 2-16: The location of the vent and intake from this Category IV appliance is regulated by the manufacturer, and this new construction may violate the required clearances.

Example 2-17: Two different style Category IV appliance vents are shown here, and each are configured in accordance with the manufacturer's requirements. The clearances around these vent terminations are also regulated by the appliance manufacturer, and due to their location at the floor rim, may likely affect the deck design.

Example 2-18: This concentric vent termination serves another Category IV appliance, whose clearances are regulated by the manufacturer.

Example 2-19: The location of this gas vent/intake is in violation of the manufacturer's minimum clearances for underneath decks.

In accordance with the manufacturer's length of vent pipe limitations, the vents were extended out to the edge of the deck.

In accordance with the manufacturer, the vent must slope upward a minimum of a 1/4 inch per 12 inches and must be insulated when in unconditioned space. Notice the deck was shimmed up by 1 1/2 inches (38 mm) from the previous photo to accommodate the required slope and insulation.

Finally, the vent terminations are installed and the installation is compliant with the manufacturer's requirements.

G2427.2.1 (503.2.3) Direct-vent appliances. Listed direct-vent *appliances* shall be installed in accordance with the manufacturer's instructions and Section G2427.8, Item 3.

G2427.8, Item 3. The vent terminal of a *direct*-vent *appliance* with an input of 10,000 *Btu* per hour (3 kW) or less shall be located at least 6 inches (152 mm) from any air opening into a building, and such an *appliance* with an input over 10,000 *Btu* per hour (3 kW) but not over 50,000 *Btu* per hour (14.7 kW) shall be installed with a 9-inch (230 mm) vent termination *clearance*, and an *appliance* with an input over 50,000 *Btu*/h (14.7 kW) shall have at least a 12-inch (305 mm) vent termination *clearance*. The bottom of the vent terminal and the air intake shall be located at least 12 inches (305 mm) above grade.

Discussion: Similar to the previous discussion, some Category I equipment utilizes a different venting method than the B-vent and single wall vent methods prescribed in the code. The most common type of direct-vent equipment is gas fireplaces (see Example 2-20). Other than Section G2427.8, Item 3, which specifies a minimum 12-inch (305 mm) clearance above grade, direct-vent equipment is regulated by the manufacturer. For deck building, in general, the only concerns are the clearances of the termination to combustibles, grade or deck below, and deck above. Clearances from building openings are also required and discussed in more detail in Chapter 8.

Example 2-20: This termination, from a direct vent fireplace, must be built around in accordance with the manufacturer's requirements and the code requirements.

Part Six: Rooftop Decks and Plumbing Vents

P3103.1 Roof extension. Open vent pipes that extend through a roof shall be terminated at least 6 inches (152 mm) above the roof or 6 inches (152 mm) above the anticipated snow accumulation, whichever is greater, except that where a roof is to be used for any purpose other than weather protection, the vent extension shall be run at least 7 feet (2134 mm) above the roof.

Discussion: The last phrases of this section are the important ones to consider when building a rooftop deck. Plumbing vent pipes are directly connected to the sewer system, and sewer gases are not a pleasant or healthy thing to have lingering around an outdoor living area. When the vents are extended 7 feet (2134 mm) above the deck, the gases can dilute into the atmosphere at a point above the occupants' heads. If the vents are not raised, and become a nuisance, an unsuspecting occupant may seal them or otherwise tamper with them (see Example 2-21). Inhibiting air movement through a plumbing vent termination can have negative results on the drain system (see Example 2-22).

Example 2-21: The two plumbing vents adjacent this rooftop living area could create quite a nuisance. Even though it's an existing situation, bidding a job to cover this roof with a decking material should include raising the sewer vent terminations.

Example 2-22: The improper use of this air-admittance valve, while eliminating odor, may negatively affect the plumbing system of the structure.

From Roof to Deck

Roofs and decks are built to very different standards that must be considered when one is to become the other. A roof framing system must be evaluated as a floor system if a deck or other type of walking surface is going to be installed on the existing structure. The dead load of the decking material may add anywhere from 4 pounds per square foot (17.78 N/.09 m^2) or more to the overall dead load of the assembly. More significantly, the live load design requirements for a roof structure may be considerably less than the live load design requirement of a floor structure, even if the roof is flat. Consider the following comparison: 2 by 6 roof rafters, #1 grade southern yellow pine, 16 inches on center with a 20 pound live load can span 15 feet 9 inches; the same criteria on floor joists with a 40 pound live load provides a maximum span of 9 feet 11 inches (see Example 2-23). This 6 foot (1829 mm) overspan would create significant deflection and could result in water collection on the roof surface or possibly complete structural failure. It's necessary to determine the type of roof construction, the materials and the location of the bearing walls below, to fully understand the capabilities of the existing roof framing system to act as a floor system.

		DEAD LOAD = 10 psf					DEAD LOAD = 20 psf				
		MAXIMUM RAFTER SPANS									
		2x4	2x6	2x8	2x10	2x12	2x4	2x6	2x8	2x10	2x12
16″	Southern Pine #1 grade	10-0	15-9	20-10	25-10		10-0	15-0	18-10	22-4	
		MAXIMUM FLOOR JOIST SPAN									
		2x4	2x6	2x8	2x10	2x12	2x4	2x6	2x8	2x10	2x12
16″	Southern Pine #1 grade	N/A	9-11	13-1	16-9	20-4	N/A	9/11	13-1	16-4	19-6

Example 2-23: These excerpts from IRC rafter and floor joist span tables display the considerable differences in allowable spans between the two construction components. The entire floor joist span table is provided in Chapter 4 of this book.

Part Seven: Exterior Veneers and Foundation Vents

R703.7.6 Weepholes. Weepholes shall be provided in the outside wythe of masonry walls at a maximum spacing of 33 inches (838 mm) on center. Weepholes shall not be less than 3/16 inch (5 mm) in diameter. Weepholes shall be located immediately above the flashing.

Discussion: It may be a surprise to you, but brick veneer on the exterior of a house is not a water-resistant barrier. Brick veneer is porous, and water, in the liquid or gas state, can be forced through the brick and mortar joints by capillary action, air pressure and temperature differences between the inside and outside of the building, and by the driving effects of a hard rain or a hose stream. Conventionally, there is an airspace behind the brick (see Example 2-24). This space and the water-resistant barrier on the outside of the wall framing create a drainage plane for liquid water to drain down the wall and exit via the weep holes at the brick ledge, just above the flashing (see Example 2-25). Most often the liquid water is collected by the condensation of warm humid air migrating toward the cooler and drier conditioned air inside the building. It's vital to maintain the originally designed escape port of this water by not blocking the weepholes. Weepholes may be in the form of tubes, wicks, holes, slots, corrugated material or any other method used to keep the opening open.

Example 2-24: The water-resistant barrier on the wall and the airspace behind the brick is designed to drain the water down the wall and out the weepholes.

Example 2-25: Blocking brick weepholes with a ledger or other building element can trap water in building cavity.

Foundation Vents

R408.1 Ventilation. The under-floor space between the bottom of the floor joists and the earth under any building (except space occupied by a *basement*) shall have ventilation openings through foundation walls or exterior walls. The minimum net area of ventilation openings shall not be less than 1 square foot (0.0929 m^2) for each 150 square feet (14 m^2) of under-floor space area, unless the ground surface is covered by a Class 1 vapor retarder material. When a Class 1 vapor retarder material is used, the minimum net area of ventilation openings shall not be less than 1 square foot (0.0929 m^2) for each 1,500 square feet (140 m^2) of under-floor space area. One such ventilating opening shall be within 3 feet (914 mm) of each corner of the building.

Discussion: If you have ever left your material or tools in the yard by the deck and covered them with a tarp for the night, you should certainly understand the thermodynamics behind this section. Water vapor from the earth is constantly moving toward the atmosphere; remember middle school science, evaporation-condensation, rain, the cycle of water on earth. This water vapor escapes the soil until it comes in contact with your tarp, chilled by the night. If the surface of the tarp is at or below the dew point of the water vapor, condensation (water) will form on the inside of the tarp. Now creatively imagine that you stored your tools in the underfloor space of a home, a crawl space. If the crawl space is like the tarp, closed to the ground without ventilation, the same condensation may occur on the underside of the floor above—obviously not a good thing. Generally, the movement of air through an underfloor space will remove the moisture-laden air prior to condensation. That's why you don't want to obstruct underfloor vents like the one in Example 2-26, but of course…they are usually located right where the ledger needs to be installed. Complete underfloor ventilation requires more than one opening, and openings near all corners. Obstructing just a couple vents could create problems.

Example 2-26: Even if a ledger was split around this vent, the deck is so low to the ground it would inhibit ventilation through the openings.

Part Eight: Safety Glazing

Definitions

DALLE GLASS. (IRC) A decorative composite glazing material made of individual pieces of glass that are embedded in a cast matrix of concrete or epoxy.

DECORATIVE GLASS. (IRC) A carved, leaded or Dalle glass or glazing material whose purpose is decorative or artistic, not functional; whose coloring, texture or other design qualities or components cannot be removed without destroying the glazing material; and whose surface, or assembly into which it is incorporated, is divided into segments.

R308.1 Identification. Except as indicated in Section R308.1.1 each pane of glazing installed in hazardous locations as defined in Section R308.4 shall be provided with a manufacturer's designation specifying who applied the designation, designating the type of glass and the safety glazing standard with which it complies, which is visible in the final installation. The designation shall be acid etched, sandblasted, ceramic-fired, laser etched, embossed, or be of a type which once applied cannot be removed without being destroyed. A *label* shall be permitted in lieu of the manufacturer's designation.

Exceptions:

1. For other than tempered glass, manufacturer's designations are not required provided the *building official* approves the use of a certificate, affidavit or other evidence confirming compliance with this code.
2. Tempered spandrel glass is permitted to be identified by the manufacturer with a removable paper designation.

Discussion: Glass, referred to by the IRC as "glazing," is an ever-present feature in practically every building in the country, yet it also provides a potential hazard to every building. This section introduces the concept of "hazardous locations" and references Section R308.4 as the method for determining their presence (see next discussion). To limit the possibility of injury or even death from large shards of broken glass, glass in hazardous locations must be safety glazing. Generally in residential applications, "tempered glass" is provided as the safety glazing standard, similar to what you would find in the rear and side windows of a truck (see Example 2-27). When shattered, tempered glass will quickly reduce to small fragments, usually less than ½ inch (12.7 mm) in cross section. These shards can still cut, but typically do not have the weight or geometry to inflict any serious injury. Glass in hazardous locations must be provided with a manufacturer's designation (a marking) providing a list of specific information. The most important in this list is the safety glazing standard to which it complies. This designation must be visible for the final inspection, and can usually be found as an etching in the corner of the glass panel. Often, due to the configuration of the interior elements of a home, some windows may already be safety glazed; locating the manufacturer's designation on the existing window will provide this evidence.

Example 2-27: Tempered glass is used in rear and side windows of automobiles for the same reason as for hazardous locations in buildings—occupant safety.

R308.4 Hazardous locations. The following shall be considered specific hazardous locations for the purposes of glazing:

1. Glazing in all fixed and operable panels of swinging, sliding and bifold doors.

   Exceptions:
   1. Glazed openings of a size through which a 3-inch diameter (76 mm) sphere is unable to pass.
   2. Decorative glazing.
2. Glazing in an individual fixed or operable panel adjacent to a door where the nearest vertical edge is within a 24-inch (610 mm) arc of the door in a closed position and whose bottom edge is less than 60 inches (1524 mm) above the floor or walking surface.

   Exceptions:
   1. Decorative glazing.
   2. When there is an intervening wall or other permanent barrier between the door and the glazing.
   3. Glazing in walls on the latch side of and perpendicular to the plane of the door in a closed position.
   4. Glazing adjacent to a door where access through the door is to a closet or storage area 3 feet (914 mm) or less in depth.
   5. Glazing that is adjacent to the fixed panel of patio doors.
3. Glazing in an individual fixed or operable panel that meets all of the following conditions:

   3.1. The exposed area of an individual pane is larger than 9 square feet (0.836 m^2); and

   3.2. The bottom edge of the glazing is less than 18 inches (457 mm) above the floor; and

   3.3. The top edge of the glazing is more than 36 inches (914 mm) above the floor; and

   3.4. One or more walking surfaces are within 36 inches (914 mm), measured horizontally and in a straight line, of the glazing.

Exceptions:

1. Decorative glazing.
2. When a horizontal rail is installed on the accessible side(s) of the glazing 34 to 38 inches (864 to 965) above the walking surface. The rail shall be capable of withstanding a horizontal load of 50 pounds per linear foot (730 N/m) without contacting the glass and be a minimum of 1 1/2 inches (38 mm) in cross sectional height.
3. Outboard panes in insulating glass units and other multiple glazed panels when the bottom edge of the glass is 25 feet (7620 mm) or more above *grade*, a roof, walking surfaces or other horizontal [within 45 degrees (0.79 rad) of horizontal] surface adjacent to the glass exterior.

4. All glazing in railings regardless of area or height above a walking surface. Included are structural baluster panels and nonstructural infill panels.
5. Glazing in enclosures for or walls facing hot tubs, whirlpools, saunas, steam rooms, bathtubs and showers where the bottom exposed edge of the glazing is less than 60 inches (1524 mm) measured vertically above any standing or walking surface.

   Exception: Glazing that is more than 60 inches (1524 mm), measured horizontally and in a straight line, from the waters edge of a hot tub, whirlpool or bathtub.
6. Glazing in walls and fences adjacent to indoor and outdoor swimming pools, hot tubs and spas where the bottom edge of the glazing is less than 60 inches (1524 mm) above a walking surface and within 60 inches (1524 mm), measured horizontally and in a straight line, of the water's edge. This shall apply to single glazing and all panes in multiple glazing.
7. Glazing adjacent to stairways, landings and ramps within 36 inches (914 mm) horizontally of a walking surface when the exposed surface of the glazing is less than 60 inches (1524 mm) above the plane of the adjacent walking surface.

   Exceptions:

   1. When a rail is installed on the accessible side(s) of the glazing 34 to 38 inches (864 to 965 mm) above the walking surface. The rail shall be capable of withstanding a horizontal load of 50 pounds per linear foot (730 N/m) without contacting the glass and be a minimum of 1 1/2 inches (38 mm) in cross sectional height.
   2. The side of the stairway has a guardrail or handrail, including balusters or in-fill panels, complying with Sections R311.7.6 and R312 and the plane of the glazing is more than 18 inches (457 mm) from the railing; or
   3. When a solid wall or panel extends from the plane of the adjacent walking surface to 34 inches (863 mm) to 36 inches (914 mm) above the walking surface and the construction at the top of that wall or panel is capable of withstanding the same horizontal load as a *guard*.
8. Glazing adjacent to stairways within 60 inches (1524 mm) horizontally of the bottom tread of a stairway in any direction when the exposed surface of the glazing is less than 60 inches (1524 mm) above the nose of the tread.

Exceptions:

1. The side of the stairway has a guardrail or handrail, including balusters or in-fill panels, complying with Sections R311.7.6 and R312 and the plane of the glass is more than 18 inches (457 mm) from the railing; or
2. When a solid wall or panel extends from the plane of the adjacent walking surface to 34 inches (864 mm) to 36 inches (914 mm) above the walking surface and the construction at the top of that wall or panel is capable of withstanding the same horizontal load as a *guard*.

Discussion: This section of the IRC provides a list of eight specific locations and criteria where the presence of glass creates an increased risk for human impact and injury. When designing a new deck, new locations for human occupancy are created. Even though the new deck may not include any glass assemblies, it will likely be adjacent glass in the existing structure, windows being the primary example. Evaluating the relationship between the existing glass and the deck design should occur during the design stage of the project, so that creation of new hazardous locations can be minimized and when created, realized. If the design must create a hazardous location, it is better to know up front that a window will need to be reglazed or otherwise protected than to find out at the final inspection or after an injury has already occurred. This discussion and the one following are only in regard to glazing in the existing structure, which may become hazardous from the addition of walking surfaces and stairways. Hazardous locations related to guard assemblies, hot tubs and doors are discussed in their respective chapters, Chapters 7 and 8.

Item 3 in the list is directly related to the location and elevation of the deck walking surface, and can affect the height at which the ledger is bolted to the structure. Any walking surface within a 36-inch (914 mm) horizontal distance from glazing may qualify as a hazardous location. A new deck could be completely separated from the existing structure and still require evaluation of the window locations (see Example 2-28). For all four listed criteria, only the area and location of each individual pane of glass is considered, not including the sash or frame of the window assembly. It is not uncommon for only specific panes in a multipane window assembly to be considered hazardous while the other may not.

Under Item 3, if the walking surface is within the 36-inch (914 mm) horizontal distance, the next criterion to consider is the size of the glass itself. Only individual panes of glass exceeding 9 square feet (0.84 m^2) in area are to be considered; however, this size limitation is only applicable to locations described in Item 3. Other hazardous locations listed may apply regardless of the glass size. The other two criteria that must apply for the location to be considered hazardous are related to the height of the top and bottom edge of the glass, measured vertically from the walking surface in question. In trying to avoid the creation of a hazardous location, the height at which the deck is constructed must be considered. The presence of a large window and the location of the deck near it may be hard to avoid, but lowering or raising the height of the deck by just a few inches can be an easy way to avoid the creation of a hazardous location (see Example 2-29). If only one of the four criteria can be avoided, the hazardous location described in Item 3 will not exist, so it is best to evaluate all options.

The first exception to the requirement for safety glazing under Item 3 is decorative glass. Decorative glass, defined earlier in this chapter, is considered less hazardous due to the small size of the individual pieces of glass and the presence of lead or other material separating and joining the pieces. Conveniently, this exception allows design flexibility in using this type of glass assembly, where safety glazing is not an option.

The second exception allows the use of a horizontal rail, intended to protect an occupant from contact with the glass. The rail, considerably small and slender in comparison to the glass panel, is not intended to block the glass completely, but rather to provide a distinct and obvious assembly to grab or fall against in the event of a slip. Notice the range of required height is identical to that of handrails. It also will minimize the possibility for an occupant to fall completely through the glass. The rail can be of any material or shape desired, provided the rail itself has a cross-sectional height of at least 1 1/2 inches (38 mm) and is capable of withstanding a 50 pound per linear foot (730 N/m) load without deflecting into contact with the glass (see Example 2-30).

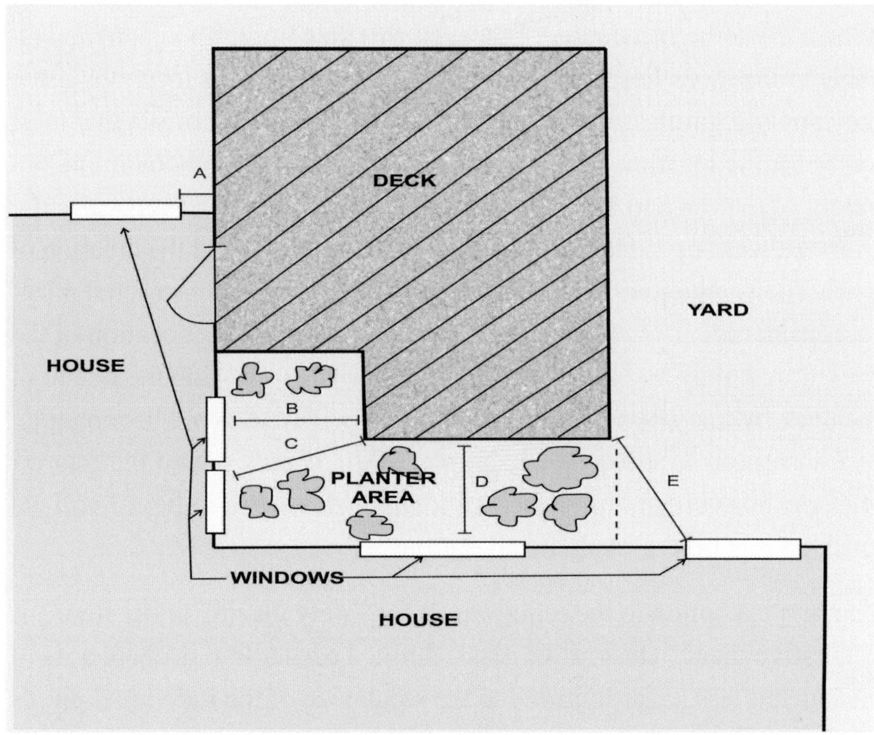

Example 2-28: The horizontal measurements labeled A, B, C, D and E in this illustration are between the deck walking surface and glazing (windows). If any of these measurements are 36 inches (914 mm) or less, an analysis of the other three criteria in Item 3 of Section R308.4 must be made.

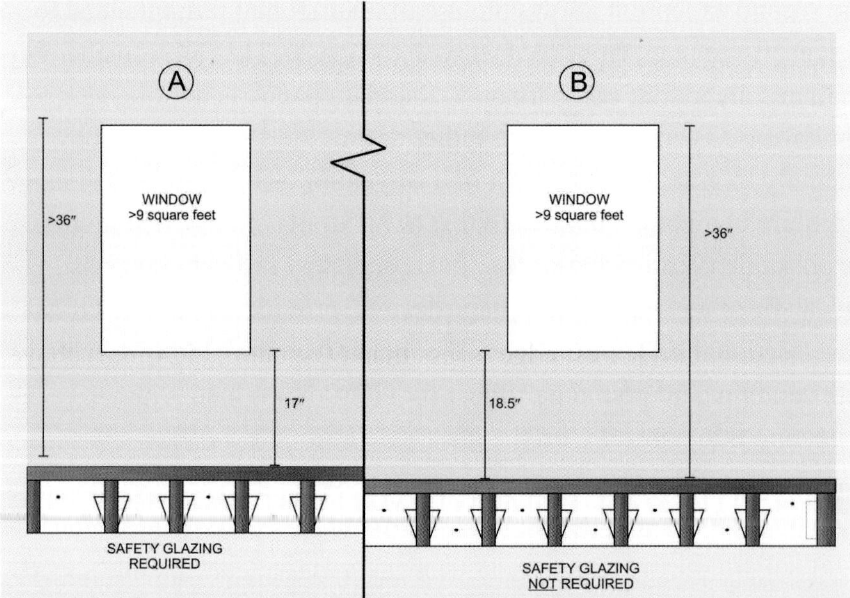

Example 2-29: In this example, a mere 1 ½ inch (38 mm) variation in the height the deck was installed is the difference between safety glazing being required and not.

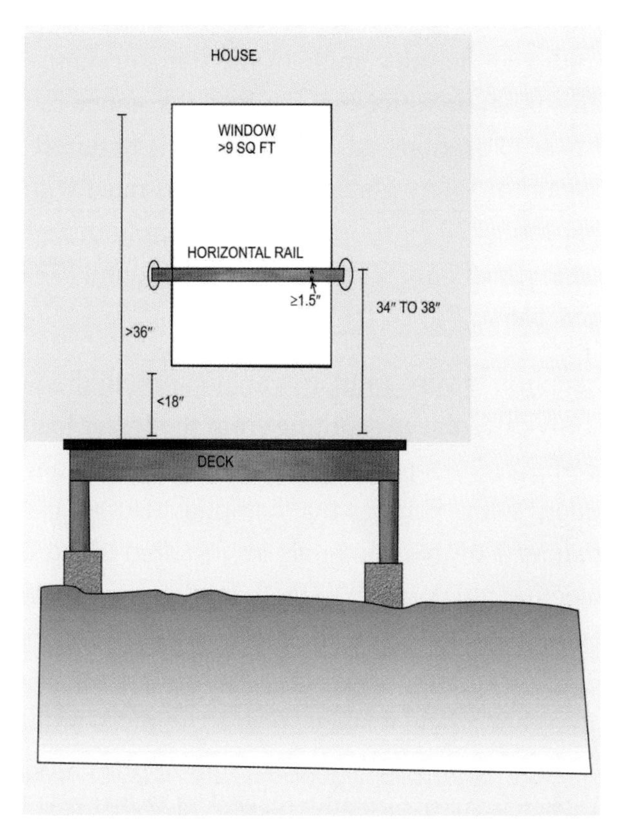

Example 2-30: Another option in lieu of safety glazing is the installation of a horizontal rail. However, this option is better if planned and discussed with the homeowners early in the project, as it will usually be visually unsatisfactory.

Discussion: Other major deck components that may become hazardous locations from their proximity to windows in the existing structure are stairways and ramps. While stumbling on a stairway is in itself a hazard, stumbling into glass is that much more so. Item 7 in the list of hazardous locations under Section R304.8 addresses this hazard specifically. Similar to walking surfaces, as previously discussed, glass within 36 inches (914 mm) of stairs, stair landings and ramps is considered a hazard if its lowest edge is less than 60 inches (1524 mm) above the walking surface of these features (see Example 2-31). When measuring the vertical height above stairs, the measurement must be taken from the nosing of the treads, just as for handrail height discussed in Chapter 7. Unlike regular walking surfaces regulated by Item 3 and previously discussed, the size of the glass area does not matter in evaluating hazardous locations under Item 7.

The first exception under Item 7 allows the installation of a protective rail identical to what was described in the exception for Item 3 (see previous discussion). However, this rail is not intended to be installed horizontally, but rather at the same angle of the stairs or ramp. It is significant at this point to notice that the load resistance requirements of the rail for these exceptions differ from what is required for stair handrails. The rail for safety glazing exemptions must withstand a 50 pound per linear foot (730 N/m) load for the entirety of its length; whereas a handrail must only resist a single 200 pound (91 kg) concentrated load (see Chapter 7).

The second exception is very similar in application to the first exception, with the only differences resulting from the lesser load resistance of a handrail compared to a protective rail, as explained above. A handrail or a guard with openings that complies with all the criteria provided in Chapter 7 of this book may be installed in the same location as the protective rail, but due to the lesser structural capabilities, they will not completely eliminate the need for safety glazing but rather reduce the 36-inch (914 mm) horizontal threshold in the criteria to an 18-inch (457 mm) horizontal distance (see Example 2-32). The third exception does, however, provide another option more similar to the protective rail when a guard assembly is provided that is solid, with no openings. The intent of this exception is to provide a short wall (a solid guard) adjacent the stairs or ramp with the top of the wall located at the same height as the protective rail, yet again with less load-resisting capabilities than the protective rail. The top of the wall will provide a noticeable and obvious ledge for people to aim their hands toward in the event of a fall and provide equivalent protection as the rail specified in the first exception.

Item 8 under the list of hazardous locations is similar to Item 7, yet is only concerned with distances from the last nosing at the bottom tread of an entire stairway. The distinction of a stair and a stairway is important when considering this location. Details of this distinction are provided in Chapter 7. When descending a stairway from one level to another, the attention of stair users will often move away from the stairway as they reach the bottom. They will begin to look toward and drift toward the direction they are going after they exit the stairway, but before reaching the final landing. If a fall were to occur from this distraction at the base of a stairway, it will likely be from the last tread, not from the landing below. This is likely why the 60-inch (1524 mm) vertical height to the bottom of a glazed assembly must be measured from the height of the lowest tread of the entire stairway, rather than from the surface of the landing (see Example 2-33). Any glazing lower than this height and within a 60-inch (1524 mm) horizontal distance from the nosing of the lowest tread must be safety glazed. It is made clear that the horizontal distance is to be measured in any direction; however, in accordance with the intent and purpose of this section, this would not usually included glazing behind the stairway.

The exceptions to this section are identical to those described in the discussion for Item 7. A protective rail, handrail or guard with openings, or a solid guard may all provide a means to limit the creation of a hazardous location.

Example 2-31: The window located adjacent to these stairs is within the range of the hazardous locations defined in Items 7 and 8 of Section R308.4, and would need to be safety glazed or provided with one of the additional protective assemblies described in the exceptions.

Example 2-32: The window adjacent this stairway is less than 60 inches (1524 mm) above the nosing of the treads and less than 36 inches (914 mm) horizontally. However, it is provided a guard that complies with the second exception under Item 7 in the list of hazardous locations, and is therefore allowed a reduced horizontal distance of 18 inches (457 mm). This stairway location and design does not require this window to be safety glazed.

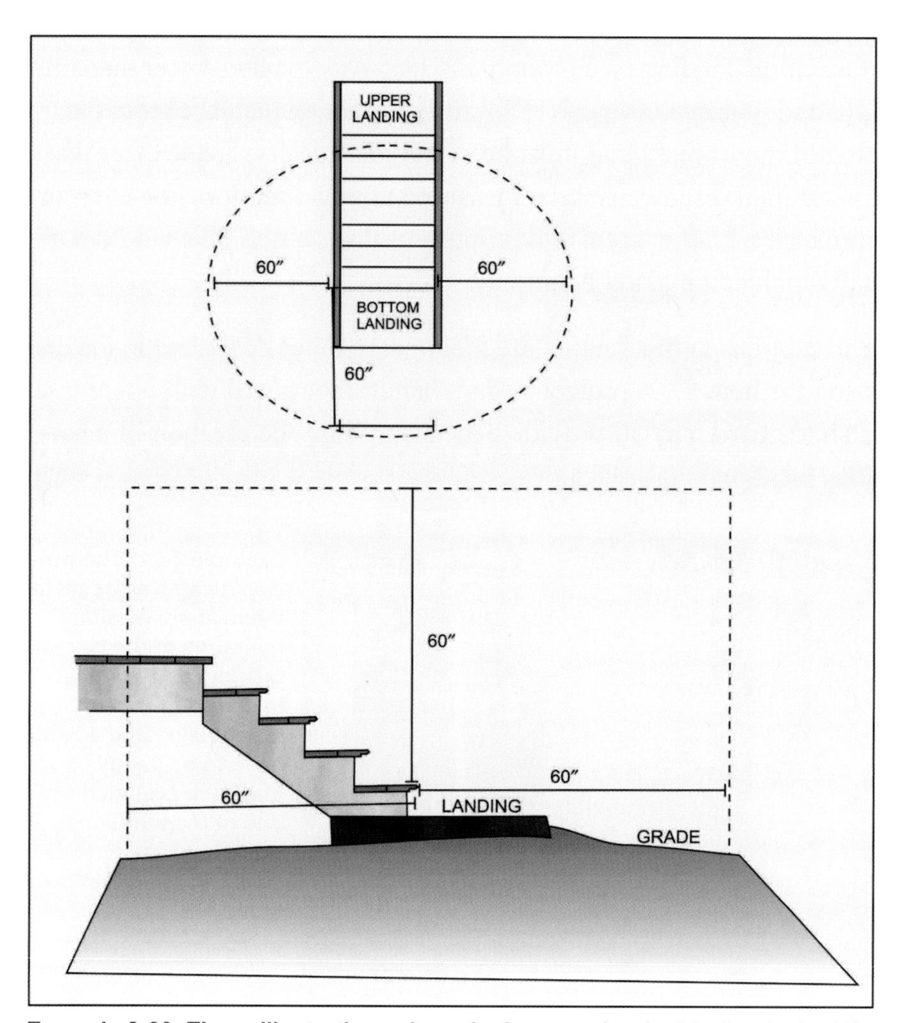

Example 2-33: These illustrations show the large region inside the dashed lines at the base of a stairway that would require safety glazing, as specified in Section R308.4, Item 8.

Part Nine: Electrical Equipment and Cables

Definitions

ACCESSIBLE, READILY (IRC). Capable of being reached quickly for operation, renewal or inspections, without requiring those to whom ready access is requisite to climb over or remove obstacles or to resort to portable ladders, etc.

DISCONNECTING MEANS (IRC). A device, or group of devices, or other means by which the conductors of a circuit can be disconnected from their source of supply.

EQUIPMENT (IRC). A general term including material, fittings, devices, appliances, luminaires, apparatus, machinery, and the like used as a part of, or in connection with, an electrical installation.

SERVICE DROP (IRC). The overhead service conductors from the last pole or other aerial support to and including the splices, if any, connecting to the service-entrance conductors at the building or other structure.

E3604.1 Clearances on buildings. Open conductors and multiconductor cables without an overall outer jacket shall have a clearance of not less than 3 feet (914 mm) from the sides of doors, porches, decks, stairs, ladders, fire escapes and balconies, and from the sides and bottom of windows that open. See Figure E3604.1.

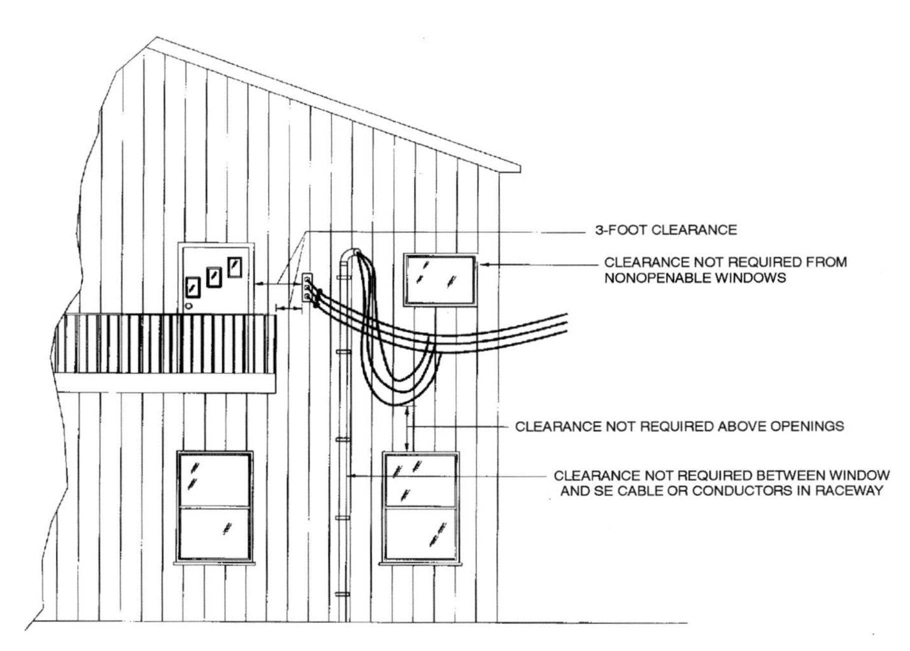

IRC Figure 3604.1: Clearances from Building Openings

E3604.2 Vertical clearances. Service-drop conductors shall not have ready access and shall comply with Sections E3604.2.1 and E3604.2.2.

E3604.2.2 Vertical clearance from grade. Service-drop conductors shall have the following minimum clearances from final grade:

1. For service-drop cables supported on and cabled together with a grounded bare messenger wire, the minimum vertical clearance shall be 10 feet (3048 mm) at the electric service entrance to buildings, at the lowest point of the drip loop of the building electric entrance, and above areas or sidewalks accessed by pedestrians only. Such clearance shall be measured from final grade or other accessible surfaces.
2. Twelve feet (3658 mm)—over residential property and driveways.
3. Eighteen feet (5486 mm)—over public streets, alleys, roads or parking areas subject to truck traffic.

Discussion: The purpose and intent of these code sections is simple: to protect people from the electrocution hazard of overhead conductors. Existing underground hazards, such as gas and electrical services, are discussed in Chapter 3. When designing a new deck, or even replacing an existing one, you must consider the location of any overhead power lines on the property and design according to the required clearances. When a deck is constructed beside an overhead service drop like that pictured in Figure 3604.1, it must be at least 3 feet (914 mm) horizontally away from most service cables. Likewise, if a deck is constructed underneath the cables, a 10 foot (3048 mm) vertical clearance between the deck and the cables must be maintained (see Example 2-34). This requirement would be the same for any "accessible pedestrian areas," such as concrete or flagstone patios, sidewalks or yards. In older homes where the cables may already be lower than 10 feet (3048 mm) over the yard, a new deck in the same spot may be prohibited by the building official.

Electrical systems and components that aren't specifically discussed in the IRC must comply with the *National Electrical Code* (NFPA 70 or NEC). Broadband cables for services such as TV and internet are regulated by Article 830 of the NEC and have similar clearances to that of electrical service cables, but are a bit lower, at 9 1/2 feet (241 mm). Technically speaking, the typical telephone lines on many homes, those that are not "network-powered broadband," are not included in any of these requirements. The clearances for these types of cables are primarily intended to separate the higher voltage electrical cables from the low-voltage communication cables, and may not be regulated during deck construction due to their limited hazard.

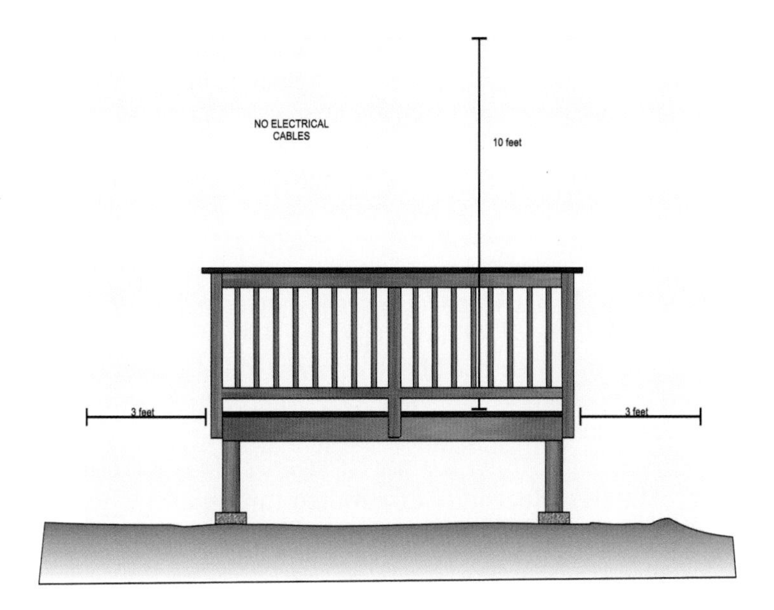

Example 2-34: A deck or other walking surface cannot be installed near overhead electrical cables, where any part of the cable is within the boundaries depicted in this illustration.

E3405.1 Working space and clearances. Sufficient access and working space shall be provided and maintained around all electrical equipment to permit ready and safe operation and maintenance of such equipment in accordance with this section and Figure E3405.1.

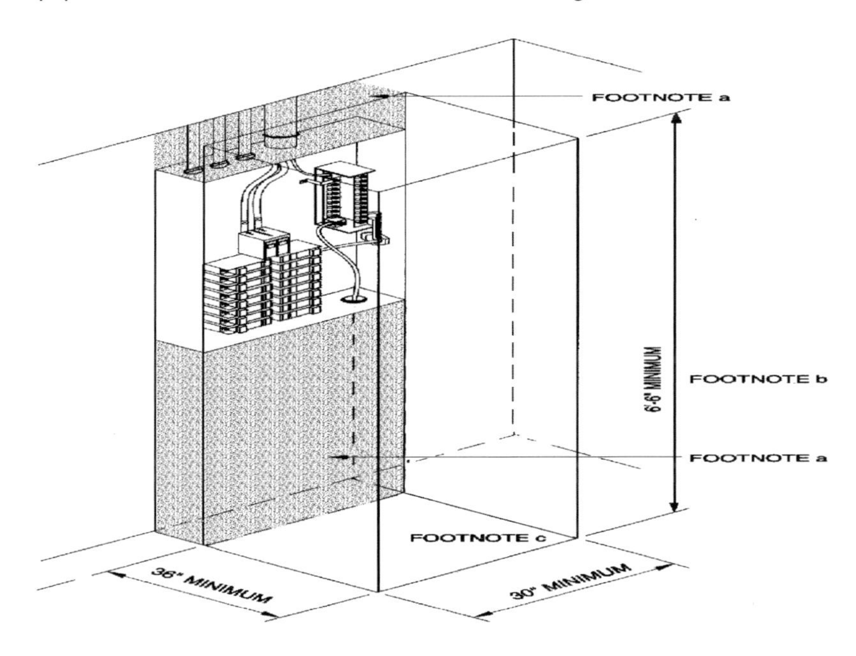

IRC Figure E3405.1: Working Space and Clearances

E3405.2 Working clearances for energized equipment and panelboards. Except as otherwise specified in Chapters 34 through 43, the dimension of the working space in the direction of access to panelboards and live parts likely to require examination, adjustment, servicing or maintenance while energized shall be not less than 36 inches (914 mm) in depth. Distances shall be measured from the energized parts where such parts are exposed or from the enclosure front or opening where such parts are enclosed. In addition to the 36-inch dimension (914 mm), the work space shall not be less than 30 inches (762 mm) wide in front of the electrical equipment and not less than the width of such equipment. The work space shall be clear and shall extend from the floor or platform to a height of 6.5 feet (1981 mm). In all cases, the work space shall allow at least a 90-degree (1.57 rad) opening of equipment doors or hinged panels. Equipment associated with the electrical installation located above or below the electrical equipment shall be permitted to extend not more than 6 inches (152 mm) beyond the front of the electrical equipment.

Discussion: The live electricity distributed throughout a home is a convenience required by the IRC, yet at the same time is one of the most potentially hazardous systems found in a home. The fire and electrocution danger of electrical systems drives the need for very stringent, cautious and specific regulations to protect the lives of the occupants, visitors and workers that may be present in or around a home. The field of electrical work is highly specialized and requires extensive experience for full comprehension of the intent and purpose of the electrical codes. By the definitions and the literal word of the IRC text, some electrical installations would be completely infeasible without a rational interpretation and approval from the local building official.

The two IRC sections provided above, E3405.1 and E3405.2, require a specifically measured working clearance for all electrical equipment. However, in accordance with the general IRC definition of "equipment," the prescribed working clearance would be completely impractical for certain types of equipment and materials. The intent of the IRC is to provide this clearance at electrical panels, meters and disconnects, and at other similar electrical equipment that may be required to be energized during repair or maintenance activities.

Various types of electrical equipment may be present at the exterior of a building that would require careful evaluation during the design stage of a deck. Relocating electrical equipment can be a very expensive chore and in most states requires the skill and expense of a licensed electrician (see Example 2-35). It is a far wiser choice to design the deck in a manner that would limit or eliminate the need for this nature of work. Main service panels and disconnects, subpanels, air-conditioning compressors and disconnects, and photovoltaic power inverters all require this prescribed working clearance (see Example 2-36).

Measured from the face of the electrical equipment, an area at least 30 inches (762 mm) wide (but not less than the width of the equipment), 36 inches (914 mm) deep and 78 inches (1981 mm) high must be provided. It is not uncommon for some allowances to be made by the local building official for specific materials and equipment. Just as a furnace or water heater is often installed and approved in a crawl space with less than 78 inches (1981 mm) of height, so might an air-conditioning compressor be allowed under a deck below this height (see Example 2-37). The intent is for safe and comfortable access for electrical personal to service the equipment, but it is the authority and interpretation of the local building official to determine how that is to be achieved (see Example 2-38).

Example 2-35: The panel to the right of the picture housed the main disconnect and the meter, yet was going to be under the deck, without the $6\frac{1}{2}$ feet (1981 mm) clearance, a clear code violation. The main disconnect was relocated to the left of the picture.

Example 2-36: Due to the location of the main panelboard, this deck was designed so that it did not encroach on the minimum 36-inch (914 mm) depth required in front of the equipment.

Example 2-37: An air-conditioning compressor and its disconnecting means are both located under this deck with less than a 78-inch (1981 mm) vertical clearance and are a violation Section E3405.2. However, in accordance with the intent and purpose of this section, a building official may allow their installation due to the relatively easy access and low hazard.

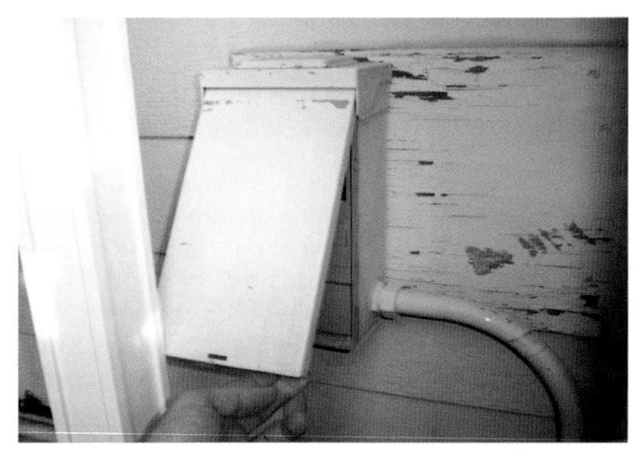

Example 2-38: Obstructing this hot-tub disconnecting means, or any electrical equipment doors, from opening at least 90 degrees (1.57 rad) is a violation of the electrical code, regardless of whether it can be de-energized or not.

E3901.7 Outdoor outlets. At least one receptacle outlet that is accessible while standing at grade level and located not more than 6 feet, 6 inches (1981 mm) above grade, shall be installed outdoors at the front and back of each dwelling unit having direct access to grade. Balconies, decks, and porches that are accessible from inside of the dwelling unit and that have a usable area of 20 square feet (1.86 m^2) or greater shall have at least one receptacle outlet installed within the perimeter of the balcony, deck, or porch. The receptacle shall be located not more than 6 feet, 6 inches (1981 mm) above the balcony, deck, or porch surface.

Discussion: The IRC requirement for outdoor receptacle outlets may appear to be solely for the convenience of the building occupant, and while it is a helpful feature, it is actually intended to limit the use and resulting fire hazard of excessive lengths of extension cords. Section E3901.7 provides two distinct requirements for outdoor receptacles. First, receptacles must be provided at grade level at the front and back of all dwellings when there is access to grade from that side of the structure. For example, if there is no back door and only a front door, only the front of the dwelling must have a grade level receptacle.

Separate from this requirement and new in the 2009 IRC, all decks, porches or balconies greater than 20 square feet (1.86 m^2) and accessible from the dwelling must be provided with a receptacle outlet within the perimeter of the space. If the dwelling has grade-level access from the side of the house where the new deck will be located, one receptacle cannot often serve both functions. For example, when an existing receptacle with grade-level access is located at one side of the dwelling and a new deck will encompass the receptacle, it will no longer satisfy its function as a grade-level receptacle, unless it can be easily accessed from the grade adjacent the deck. The existing receptacle may now satisfy the requirement for the deck, but an additional one must be provided that is accessible from grade. If a new deck is constructed that will not encompass the existing grade-level receptacle or one did not previously exist, only the receptacle required for the deck must be installed (see Example 2-39).

The implications of this code modification in the 2009 IRC will almost always require the installation of a new receptacle outlet when a deck, patio or balcony is constructed that is accessible from the dwelling. In installing new electrical outlets, there are other requirements beyond basic wiring techniques, such as tamper-resistant receptacles and in-use covers, all of which are outside of the scope of this book. It is always suggested and often required by the state authority to employ the expertise of a licensed electrician when performing alterations to electrical systems.

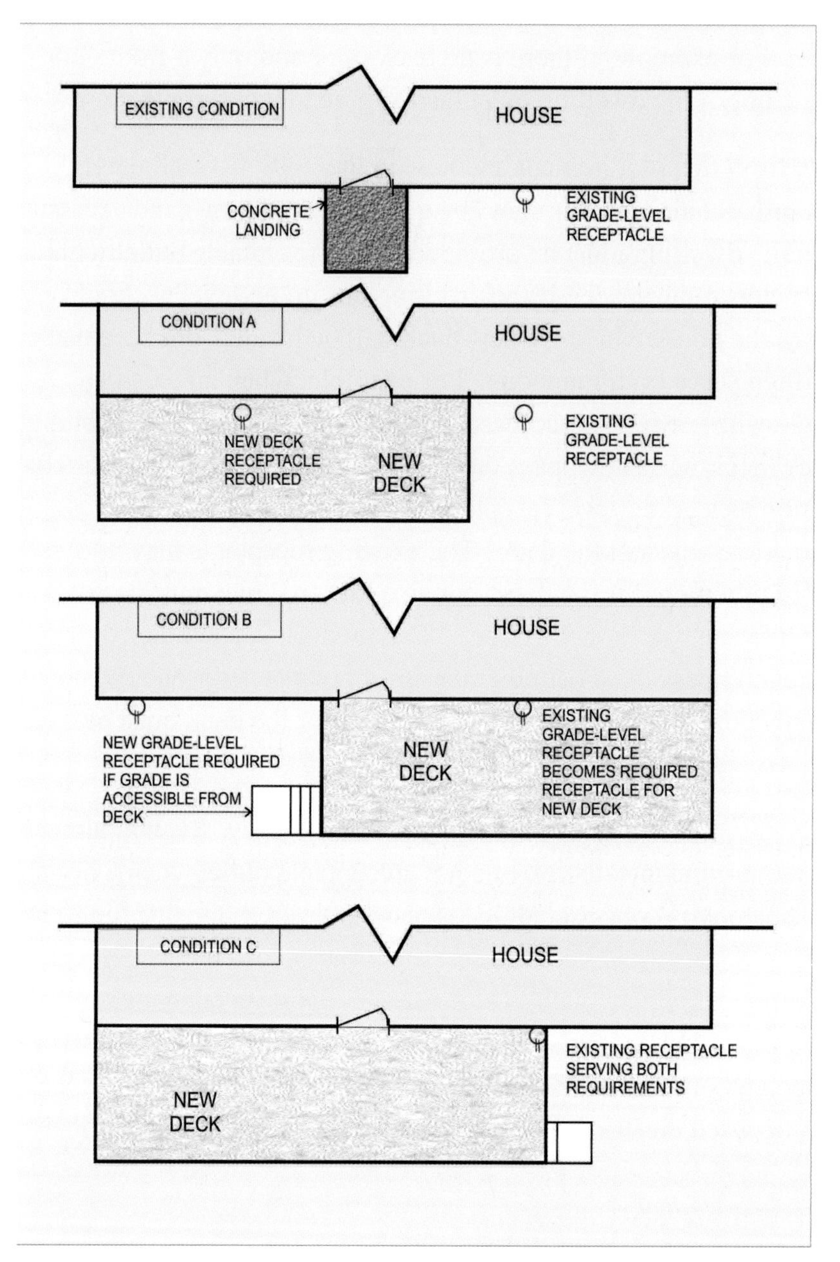

Example 2-39: In the existing condition, a back door of the dwelling accesses grade level and is already provided an existing grade-level accessible outlet. In condition "A," a new deck is constructed and an additional receptacle must be installed to serve the deck. In condition "B," the new deck was constructed around the existing grade-level accessible receptacle, and this receptacle now acts as the receptacle required to serve the deck. If the new deck provides access to grade, an additional receptacle will be required to replace the previous grade-level receptacle, but if the new deck is surrounded by guards or does not have a compliant stairway to access grade level an additional receptacle is not required. In condition "C," the existing receptacle is at the edge of the deck and is able to serve both the deck and be accessible from grade level. In this condition no additional receptacles are required.

Part Ten: Emergency Escape and Rescue Openings

Definitions

COURT (IRC). A space, open and unobstructed to the sky, located at or above *grade* level on a *lot* and bounded on three or more sides by walls or a building.

EMERGENCY ESCAPE AND RESCUE OPENING (IRC). An operable exterior window, door or similar device that provides for a means of escape and access for rescue in the event of an emergency.

YARD (IRC). An open space, other than a court, unobstructed from the ground to the sky, except where specifically provided by this code, on the *lot* on which a building is situated.

R310.1 Emergency escape and rescue required. *Basements*, habitable attics and every sleeping room shall have at least one operable emergency escape and rescue opening. Where *basements* contain one or more sleeping rooms, emergency egress and rescue openings shall be required in each sleeping room. Where emergency escape and rescue openings are provided they shall have a sill height of not more than 44 inches (1118 mm) above the floor. Where a door opening having a threshold below the adjacent ground elevation serves as an emergency escape and rescue opening and is provided with a bulkhead enclosure, the bulkhead enclosure shall comply with Section R310.3. The net clear opening dimensions required by this section shall be obtained by the normal operation of the emergency escape and rescue opening from the inside. Emergency escape and rescue openings with a finished sill height below the adjacent ground elevation shall be provided with a window well in accordance with Section R310.2. Emergency escape and rescue openings shall open directly into a public way, or to a *yard* or court that opens to a public way.

> **Exception:** *Basements* used only to house mechanical *equipment* and not exceeding total floor area of 200 square feet (18.58 m^2).

R310.1.1 Minimum opening area. All emergency escape and rescue openings shall have a minimum net clear opening of 5.7 square feet (0.530 m^2).

> **Exception:** *Grade* floor openings shall have a minimum net clear opening of 5 square feet (0.465 m^2).

R310.1.2 Minimum opening height. The minimum net clear opening height shall be 24 inches (610 mm).

R310.1.3 Minimum opening width. The minimum net clear opening width shall be 20 inches (508 mm).

R310.1.4 Operational constraints. Emergency escape and rescue openings shall be operational from the inside of the room without the use of keys, tools or special knowledge.

Discussion: IRC Sections R310.1 through R310.1.4 provide very important and specific life-safety requirements for emergency escape and rescue from dwellings. In the event of a fire or other emergency in a home, the greatest concern is placed on individuals who may be asleep during the emergency. (I am sure you know someone that sleeps like a rock and is not easily shaken awake. These folks, as well as the rest of us, will experience a delayed reaction to a fire emergency.) During this delay, the normal means of exiting the building may already be blocked and unusable. It is this reason that the IRC requires all "sleeping rooms" to contain an emergency escape and rescue opening that leads directly to a public way, or to a yard or court that leads to a public way. This is not meant to imply that the opening cannot lead onto a deck, roof or other element of the building. Often basements, even those that do not contain sleeping rooms, are used as sleeping rooms for certain occasions. Consider where children's slumber parties or out-of-town guests are often housed. Because of this common practice, all basements (finished or not) are also required to have one of these openings. The only exception is basements not exceeding 200 square feet (18.5 m^2) and used only to house mechanical equipment, such as space and water heating appliances.

While the emergency escape and rescue opening can be a full-size door, it is most often a window, and is commonly referred to as an "egress window." Regardless of this commonly used term, the minimum horizontal and vertical dimensions as well as the overall minimum open area are actually based on the "rescue" part of the requirement. These dimensions have been researched by fire protection agencies and found to be the minimum size necessary for rescue personal in full gear to safely gain access through the opening.

While deck construction does not usually involve window modifications, it may inhibit the requirement for the opening to lead to a public way. When designing elements that affect the exterior of a home, locating the sleeping rooms is of extreme importance. Constructing a deck that inhibits escape or rescue from a required opening could leave you with a haunted conscience and a costly liability in the event of a fire.

The dimensions of these openings have not always existed in the code in this specific way, so smaller windows may be the only opening from a bedroom. Regardless of the history of the dwelling or the size of the windows, the rooms used as sleeping rooms must have a clear path from the building from whatever window is closest to compliance—you may still be able to get out a smaller, noncompliant window, but not with a deck in the way. Similarly, if the basement contains no bedrooms and does not have a compliant opening, it would be a wise choice to leave a clear path from the basement's largest and most accessible window opening. The following sections, R310.2, R310.2.1, R310.4 and R310.5, contain specific provisions to ensure an unobstructed path from the outside of the opening to a point clear from the building.

R310.2 Window wells. The minimum horizontal area of the window well shall be 9 square feet (0.9 m^2), with a minimum horizontal projection and width of 36 inches (914 mm). The area of the window well shall allow the emergency escape and rescue opening to be fully opened.

> **Exception:** The ladder or steps required by Section R310.2.1 shall be permitted to encroach a maximum of 6 inches (152 mm) into the required dimensions of the window well.

Discussion: The current dimensions required for window wells serving emergency escape and rescue openings with sills below finished grade are specified in this section. In the event that you are unable to determine the location of the bedrooms, or there are no bedrooms in the basement, one clue may be the size of the exterior window wells. Often in new construction, home builders will only provide one window well that is 36 inches (914 mm) in depth, with the remaining ones being only 30 inches (762 mm) (see Example 2-40). The single 36 inches (914 mm) well is intended to serve the opening required for the basement and needs to be considered during the design process.

Example 2-40: The deeper window well is the single required emergency escape and rescue opening provided to serve the unfinished basement, and consideration must be taken in designing a deck around or above it.

R310.2.1 Ladder and steps. Window wells with a vertical depth greater than 44 inches (1118 mm) shall be equipped with a permanently affixed ladder or steps usable with the window in the fully open position. Ladders or steps required by this section shall not be required to comply with Sections R311.7 and R311.8. Ladders or rungs shall have an inside width of at least 12 inches (305 mm), shall project at least 3 inches (76 mm) from the wall and shall be spaced not more than 18 inches (457 mm) on center vertically for the full height of the window well.

Discussion: For decks constructed close to the ground, one method of maintaining a path from the window well away from the building is to build the deck around the window well, or leave an opening in the deck if above the top of the window well. One consideration with this design is the need for a permanent ladder for vertical wells deeper than 44 inches (1118 mm). If the occupant is required to reach the deck surface rather than just the top of the window well, a ladder may need to be installed or extended, such that it meets the requirements specified in this section (see Example 2-41).

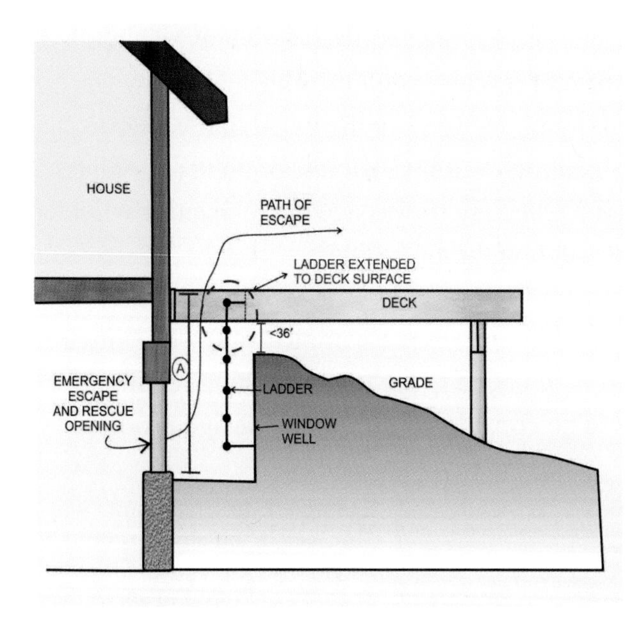

Example 2-41: If a 36-inch (914 mm) high path cannot be provided beneath a deck from an emergency escape and rescue opening (see Section R310.5), the path can be brought through and onto the deck surface. The deck would be considered the top of the window well, and if the new depth of the window well, measured from the deck surface to the bottom of the well, exceeds 44 inches (1118 mm) a ladder would need to be provided or extended.

R310.4 Bars, grilles, covers and screens. Bars, grilles, covers, screens or similar devices are permitted to be placed over emergency escape and rescue openings, bulkhead enclosures, or window wells that serve such openings, provided the minimum net clear opening size complies with Sections R310.1.1 to R310.1.3, and such devices shall be releasable or removable from the inside without the use of a key, tool, special knowledge or force greater than that which is required for normal operation of the escape and rescue opening.

Discussion: When decks are designed as previously discussed, with an opening in the deck surface for the window well, a new issue is created—deep holes in the deck surface create quite a fall hazard. A common practice in these situations is the construction of removable panels of decking material that serve to cover the opening and keep the deck safe, usable and attractive. While this certainly provides positive results for the deck, it creates terribly dangerous or even deadly results to the emergency escape and rescue operation. The IRC does allow the emergency escape and rescue opening window wells to be covered, but within some limiting parameters.

The minimum size of 9 square feet (0.9 m^2) specified in Section R310.2 must be maintained at the opening of the cover, and considering the minimum required 36 inches (914 mm) depth, the opening and cover must be at least 36 inches by 36 inches (914 mm by 914 mm). The cover cannot be secured in place by any means that requires a key, tool or special knowledge. While considering what "special knowledge" means, think about the purpose for this opening—to get people out in an emergency. Now consider what you would be able to do if suddenly awoken at 3 am by the screech of smoke alarms to a dark room and eyes burning from smoke. Panicking and half-asleep, in these conditions what may have seemed to be a simple procedure for unlatching the window well cover may no longer be so simple…it may feel like "special knowledge." The safest and most foolproof approach is the absence of any latch and removal of the cover with a simple PUSH!

The final sentence of this section is concerned directly about that "push" and just how hard it must be. Removing the window well cover cannot require a force greater than that required to operate the emergency escape and rescue opening. It is not likely that the window would take as much force to operate as lifting a 9-square-foot (0.9 m^2) area of decking material. This IRC section would make it very difficult, if not impossible, to construct a window well cover of typical decking material (see Example 2-42).

While rescue is an essential function of the emergency escape and "rescue" openings, the code does not require window well covers to be removable from the exterior. In the balance of allowing entry for rescue and inhibiting entry for security, security weighs heavier. Regardless of the minimum requirements of the IRC, discussing these issues with the homeowner is highly suggested.

Example 2-42: This removable deck section over an emergency escape and rescue opening is not deep enough, and is likely already too heavy to comply with the IRC.

R310.5 Emergency escape windows under decks and porches. Emergency escape windows are allowed to be installed under decks and porches provided the location of the deck allows the emergency escape window to be fully opened and provides a path not less than 36 inches (914 mm) in height to a *yard* or court.

Discussion: There are basically two options to designing a deck around or over a window well serving an emergency escape and rescue opening; the first option was discussed previously. You can build the deck over the well and provide an opening in the deck or a lightweight cover directly above the top of the well, and extend the ladder up to the top of the deck surface if required (see Example 2-43). Another option, if the deck is high enough, is to provide a passage beneath the deck to exit from the window well to a yard or court and finally to a public way. The path must be at least 36 inches (914 mm) high measured from the top of the well opening and from grade for the length of the path. There is no minimum width required, but most likely a building official would require at least 36 inches (914 mm) of width, as this is a common dimension for other exit components such as stairs and hallways (see Example 2-43).

Example 2-43: This window is serving as an emergency escape and rescue opening for a lower level bedroom. The deck provides a 36-inch (914 mm) high clear path out from under the deck, and the lattice skirting was stopped short of the house to further provide a clear path of egress.

Chapter 3: Foundations

Introduction

The final resting place for all the forces imposed on a structure is the earth, and the foundation system is the final means to deliver these forces. The predominant direction of these forces is downward, toward the earth, and they are absorbed and resisted by the supporting soil beneath the foundation system. While relatively minimal for basic deck construction, lateral and uplift forces may also be imposed on a foundation system; however, they are usually reduced or eliminated by the counteracting weight (dead load) of the structure itself. A belled pier or footing projection can resist these uplift forces, as well as the friction between the sides of the foundation and the surrounding soil as a result of the soil's lateral pressure against the foundation. In cold climates where the earth is subject to frost heave, too much friction, from uneven and rough edges of the excavation, can pull the pier up by the adjacent heaving soil. Some forces imposed on a structure originate in the earth and end at the earth. Seismic forces from the ground shake and rattle a structure as a whole, but that shake is still resisted by the more localized soil adjacent the foundation system.

When a foundation is not designed with regard to the properties of the soil beneath and adjacent the foundation system, the ability for the soil to resist the forces can be compromised. Soil that is compressible, unstable, expansive, loose, subject to freeze/thaw cycles or containing an excessive amount of organic material or large debris can inhibit the function of the foundation system.

In general, the support posts for residential decks, whether wood, concrete or steel, will typically bear on pier foundations constructed of concrete, masonry, treated wood or other appropriate material. For example, a concrete pier constructed using a cardboard forming tube and concrete poured inside may rest on a concrete pad footing at the bottom of the hole. If properly sized at the bottom to carry the deck load, a pier foundation may not actually need a separate concrete pad footing at the bottom. The bottom of these foundations will need to extend downward to the designated frost depth. In order to properly size the footing area at the base of the pier footing/foundation, a determination needs to be made about the soil bearing capacity. There will be a discussion of this issue later in this chapter. If the soil bearing capacity cannot be determined, the local building official has the authority to require an evaluation provided by a registered engineer competent in soil mechanics.

Part One: Soil-Bearing Capacity

Definitions

Bearing Capacity (Load-bearing pressure). (McGraw-Hill) Load per unit area which can be safely supported by the ground

Undisturbed [soil]. (McGraw-Hill) Pertaining to a sample of material, as of soil, subjected to so little disturbance that it is suitable for determinations of strength, consolidation, permeability characteristics, and other properties of the material in place.

R401.4 Soil tests. Where quantifiable data created by accepted soil science methodologies indicate expansive, compressible, shifting or other questionable soil characteristics are likely to be present, the *building official* shall determine whether to require a soil test to determine the soil's characteristics at a particular location. This test shall be done by an *approved agency* using an *approved* method.

R401.4.1 Geotechnical evaluation. In lieu of a complete geotechnical evaluation, the load-bearing values in Table R401.4.1 shall be assumed.

TABLE R401.4.1
PRESUMPTIVE LOAD-BEARING VALUES OF FOUNDATION MATERIALS[a]

CLASS OF MATERIAL	LOAD-BEARING PRESSURE (pounds per square foot)
Crystalline bedrock	12,000
Sedimentary and foliated rock	4,000
Sandy gravel and/or gravel (GW and GP)	3,000
Sand, silty sand, clayey sand, silty gravel and clayey gravel (SW, SP, SM, SC, GM and GC)	2,000
Clay, sandy clay, silty clay, clayey silt, silt and sandy silt (CL, ML, MH and CH)	1,500

For SI: 1 pound per square foot = 0.0479 kPa.
a. When soil tests are required by Section R401.4, the allowable bearing capacities of the soil shall be part of the recommendations.
b. Where the building official determines that in-place soils with an allowable bearing capacity of less than 1,500 psf are likely to be present at the site, the allowable bearing capacity shall be determined by a soils investigation.

Discussion: The bearing area of a foundation system is the only means allowed, prescriptively, by the IRC for transmitting the downward loads imposed on a structure to the earth. This area is the horizontal area at the base of a footing or pier that applies a compressive pressure against the soil. The larger the bearing area, the greater loads the soil can resist; a larger surface area distributes the load over more soil. However, bearing area is not the only piece of the load-transmitting equation. The type of soil under the foundation may vary in stability and load-bearing capacity. An equivalent bearing area on differing soil types will yield different total bearing capacities, in the same manner that equivalent soil types in contact with different size bearing areas will produce different bearing capacities.

Before a foundation system can be sized and designed, the tributary load imposed on each isolated footing or pier must be determined (see Example 3-1). Once this load is determined for each footing, the soil bearing capacity must be determined. Numerous types of soil are provided in Table 401.4.1, referenced by the above IRC section, along with their allowable bearing capacity in pounds per square foot of bearing area. However, determining which type of soil is present is not easily definitive. Sandy silt and silty sand may be hard to distinguish if you are not a geotechnical professional, yet these two soil types have different bearing capacities. In order to simplify deck permit approval, the local building official, based on his or her knowledge of the region, may have predetermined the predominant soil type in the region. Other times he or she may simply default to the lowest soil bearing capacity in Table 401.4.1, 1,500 pounds per square foot. In the absence of simplified or predetermined soil capacities such as these, the building official may require a geotechnical evaluation (soils analysis) for the purpose of calculating the required bearing area. The method required to determine the soil bearing capacity is at the discretion of the building official, and may vary from project to project depending on the magnitude of the tributary loads, design of the deck or the availability of records of the soil analysis performed when the existing structure was constructed.

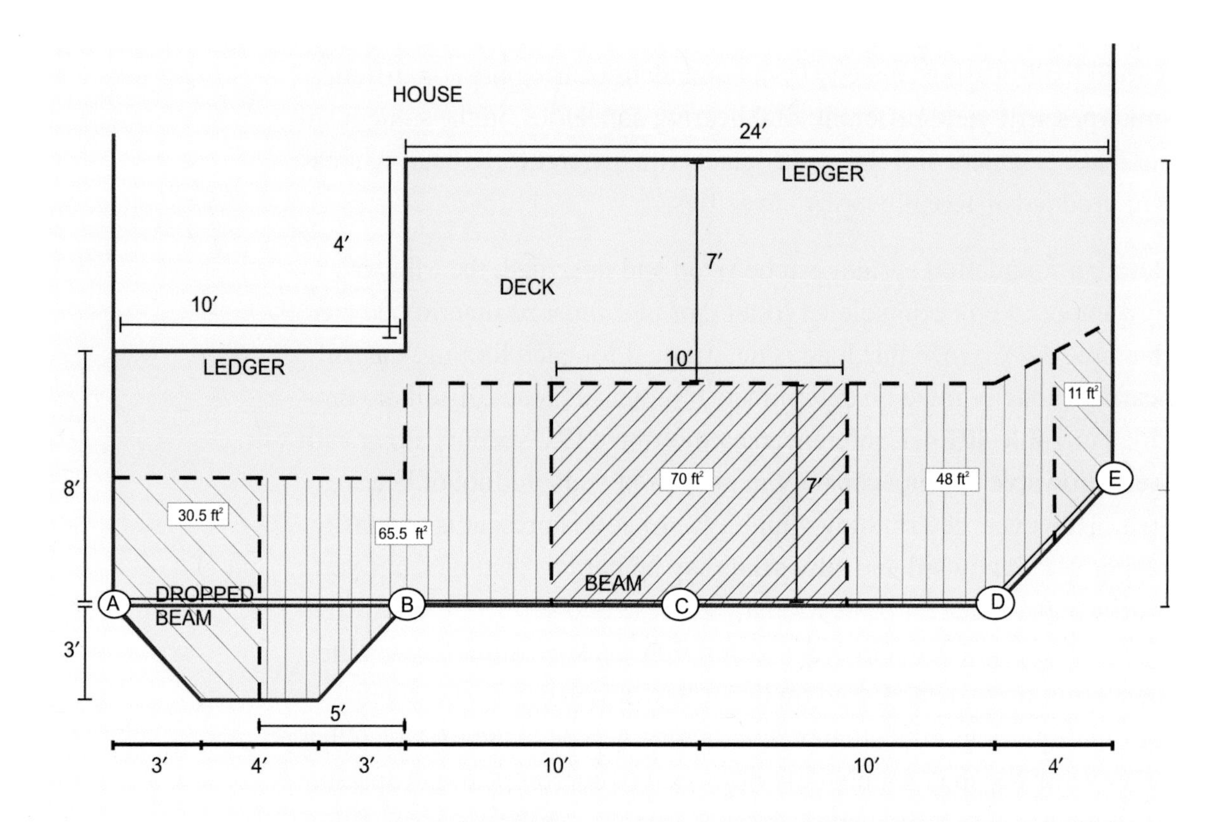

Example 3-1: To determine the load on each deck post, the area of the deck supported by each must be determined. For basic designs not supporting concentrated loads, a simple method of dividing the spans can be used. First the beam spans are divided in half, then the joist spans. Notice the significant variation in the loads supported in this example, from 11 square feet (1.02 m²) to 70 square feet (6.5 m²). For simplicity and uniformity in the finished project, all the footings could be sized based on the largest load, or each footing could be sized individually. The unshaded areas of the deck in this example are supported by the existing structure through the ledger connection.

R403.1.4 Minimum depth. All exterior footings shall be placed at least 12 inches (305 mm) below the undisturbed ground surface. Where applicable, the depth of footings shall also conform to Sections R403.1.4.1 through R403.1.4.2.

Discussion: It is important to recognize in this section that the minimum depth must be measured from the top of the "undisturbed ground surface," not the top of the final grade. This depth is the minimum required and applies to all foundation systems; however, more specific requirements in other sections, such as frost protection, may require an even greater depth. Bearing at a depth 12 inches (305 mm) below the undisturbed soil will provide a more stable surface, less likely to settle or shift than disturbed soil. Lateral support and resistance to overturning will also be provided by the undisturbed soil at the sides of the foundation. Precast concrete footings are permitted by the IRC, but still must comply with all the provisions of cast-in-place foundations (see Example 3-2).

Determining what is and is not undisturbed soil may not be an easy or practical task, and often a few assumptions may be made by the local building official. The soil adjacent to the foundation of the existing structure has in most cases been disturbed during construction (see Example 3-3); an exception to this being a cast-in-place foundation system that does not require overexcavation and backfill practices. Depending on the foundation type and depth of the existing structure, the area of disturbed soil may extend 4 to 5 feet (1219 to 1524 mm) horizontally from the existing foundation wall and as deep as the existing foundation wall. Once the earth has been excavated, the soil can never again be considered "undisturbed," unless filled, compacted and tested through an engineering process (see Example 3-4). Generally, it's a good idea to keep deck foundations at least 5 feet (1524 mm) from existing foundations, or excavate to the depth of the existing foundation, where undisturbed soil is located. In other areas of a property, undisturbed soil may also be below the final grade, such as in flower beds, yards, soil embankments and fill at retaining walls (see Example 3-5).

As discussed in Chapter 1, alternative methods that would perform in an equivalent manner to the prescriptive methods of the code can be approved. One such way to avoid the difficulties of backfill instability against an existing foundation is to use the existing foundation (see Example 3-6).

Side Note: When encountering a cantilever, brick veneer or other condition of the existing structure that prohibits connection of a typical ledger, attention must be given to this IRC section, as it restricts the location of posts for free-standing decks.

Example 3-2: Precast concrete footings can be used, but must bear at the required depth on suitable, undisturbed soil. This installation will not provide a sufficient foundation for any structure.

Example 3-3: The backfill soil that will be placed against this foundation after the forms are removed cannot be considered "undisturbed" and is inadequate for bearing capacity.

Example 3-4: The large void underneath this concrete landing is the result of unstable backfill that has been settling over the years. The uncertainty in determining whether foundation backfill has completely stabilized and compacted is why it is rarely considered "undisturbed" soil.

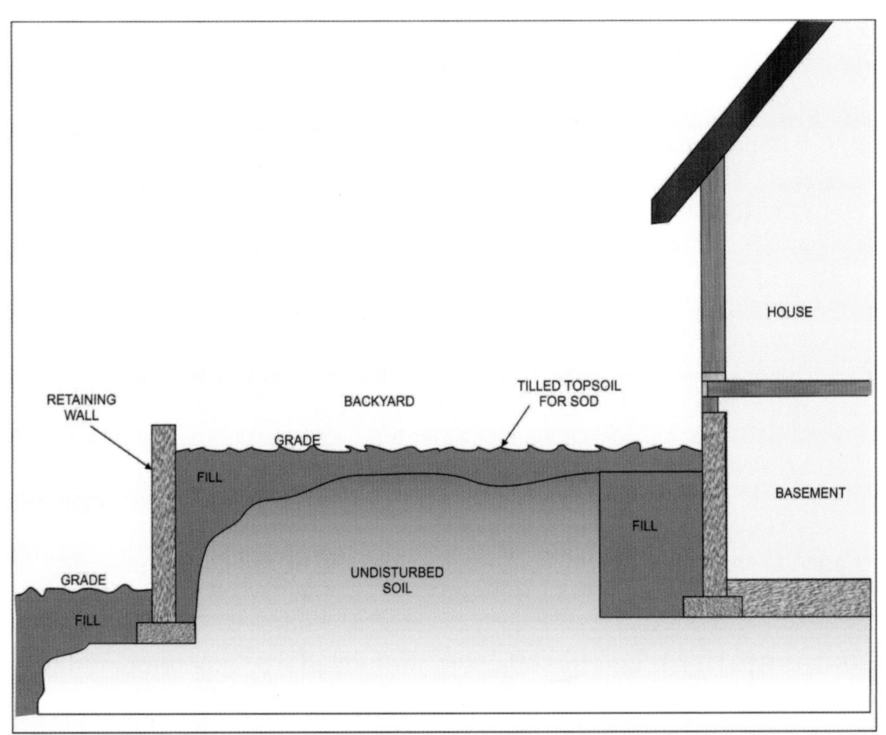

Example 3-5: Only once the undisturbed soil has been reached can the depth of the excavation begin to be measured.

Example 3-6: Rather than attaching the deck to the brick veneer and violating IRC Section R703.7.3 (see Chapter 4), this free-standing deck used a beam and posts in lieu of a ledger. To avoid the issues of bearing a deck footing in backfill, an engineer-designed bracket was bolted to the existing foundation. This design must be approved as an alternative.

R403.1.4.1 Frost protection. Except where otherwise protected from frost, foundation walls, piers and other permanent supports of buildings and structures shall be protected from frost by one or more of the following methods:

1. Extended below the frost line specified in Table R301.2(1);
2. Constructing in accordance with Section R403.3;
3. Constructing in accordance with ASCE 32; or
4. Erected on solid rock.

Exceptions:

1. Protection of freestanding *accessory structures* with an area of 600 square feet (56 m^2) or less, of light-framed construction, with an eave height of 10 feet (3048 mm) or less shall not be required.
2. Protection of freestanding *accessory structures* with an area of 400 square feet (37 m^2) or less, of other than light-framed construction, with an eave height of 10 feet (3048 mm) or less shall not be required.
3. Decks not supported by a dwelling need not be provided with footings that extend below the frost line.

Footings shall not bear on frozen soil unless the frozen condition is permanent.

Discussion: In regions subject to prolonged freezing temperatures, the moisture within the soil may freeze. When the soil beneath a foundation system freezes, its expansion can subject the foundation system to forces, pressures and movement it has not been designed to accommodate. For foundations with minimal loads imposed, as often occurs in deck construction, the expansion of the soil can cause the foundation to heave upward. When thawing occurs and the expansion of the moist soil subsides, the void left beneath the foundation can fill with water-transported sediment and silt. Filling in this newly created void can inhibit the foundation from settling back to its original depth, and it may alter or diminish the bearing properties of the soil.

Two of the IRC methods for foundation frost protection are applicable to decks. The first, and most common, is simply excavating the foundation system to below the local frost line. The frost line is determined by the local building official, based on local experience, the US Weather Bureau, or any other source deemed acceptable to the building official. Table R301.2(1), provided and discussed in Chapter 4, provides a fill-in-the-blank location for the local jurisdiction to officially document the climatic and geographic design criteria they have established for their region.

Often, excavating to the depth of the frost line can be inhibited by low-lying subsurface rocks. In the event that solid rock is discovered prior to reaching frost depth, the IRC recognizes it as suitable for frost protection. The intention is that foundations supported directly on solid rock are not susceptible to frost action and thus will be protected.

A deck-specific exception for frost-protected foundations is provided in this section. A free-standing deck that is not supported by the primary structure does not require frost protection of the foundation. Presumably, the foundation system of an existing structure is frost protected, and supporting a deck partially from that foundation and partially from another, without frost protection, can create differential movement across the deck structure. If this occurs, the connections between various structural members, such as joists, beams and ledgers, can be compromised by abnormal stress and deformation that has not been considered in the design. A free-standing deck supported by a single foundation system of isolated piers without frost protection will theoretically experience movement as a whole, which may help to maintain the integrity of the structural connections in the deck. In a multilevel deck, where perhaps an upper deck is connected to the existing structure by a ledger and a lower deck is provided with piers on all sides, each deck platform is considered a separate deck. In these designs, the piers supporting the ledger-connected deck would be the only ones requiring frost protection, provided the decks do not share common piers or structural connections (see Example 3-7).

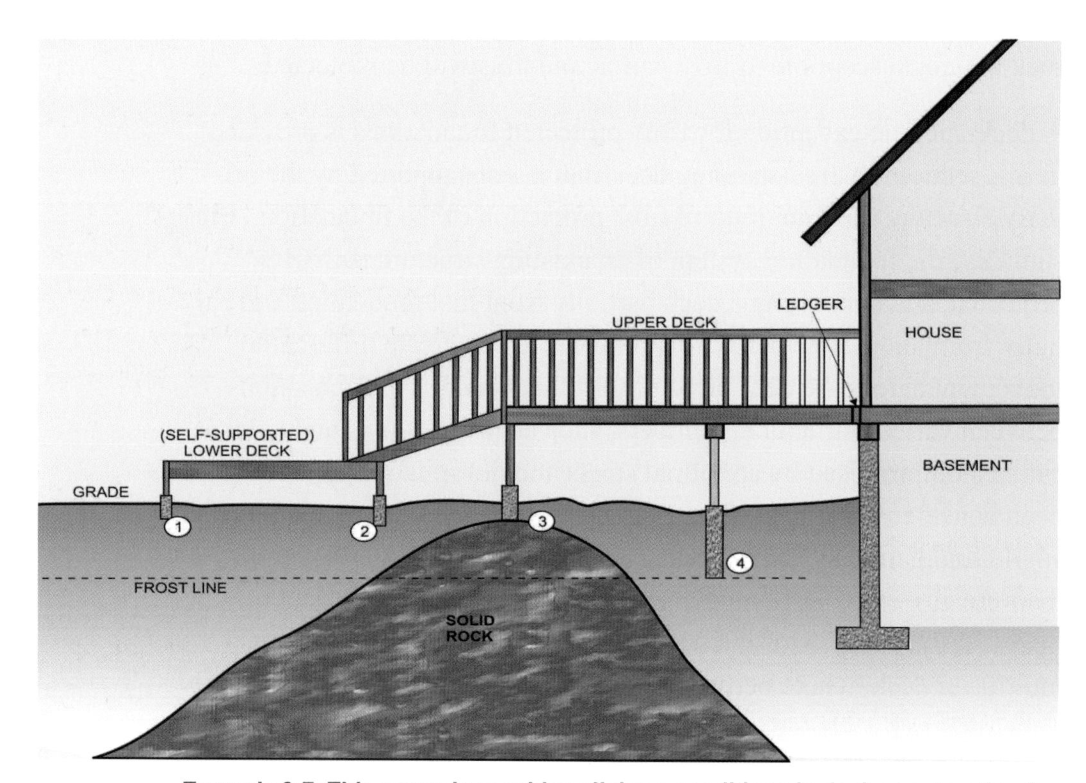

Example 3-7: This example provides all three conditions in deck construction for frost protection of foundations. Piers 1 and 2 do not have to reach frost depth as they support a free-standing deck. Piers 3 and 4 are required to reach frost depth, as the deck is also supported by the existing structure. Pier 3, however, satisfies the frost depth requirement by bearing on solid rock. Piers 1 and 2 must satisfy the general requirement of bearing 12 inches (305 mm) below undisturbed ground level.

Part Two: Foundation Design

Definitions

Footing. (McGraw-Hill) The widened base of substructure forming the foundation for a wall or a column.

R403.1 General. All exterior walls shall be supported on continuous solid or fully grouted masonry or concrete footings, crushed stone footings, wood foundations, or other *approved* structural systems which shall be of sufficient design to accommodate all loads according to Section R301 and to transmit the resulting loads to the soil within the limitations as determined from the character of the soil. Footings shall be supported on undisturbed natural soils or engineered fill. Concrete footing shall be designed and constructed in accordance with the provisions of Section R403 or in accordance with ACI 332.

R403.1.1 Minimum size. Minimum sizes for concrete and masonry footings shall be as set forth in Table R403.1 and Figure R403.1(1). The footing width, *W*, shall be based on the load-bearing value of the soil in accordance with Table R401.4.1. Spread footings shall be at least 6 inches (152 mm) in thickness, *T*. Footing projections, *P*, shall be at least 2 inches (51 mm) and shall not exceed the thickness of the footing. The size of footings supporting piers and columns shall be based on the tributary load and allowable soil pressure in accordance with Table R401.4.1. Footings for wood foundations shall be in accordance with the details set forth in Section R403.2, and Figures R403.1(2) and R403.1(3).

R403.1.3 Seismic reinforcing.

Exception: In detached one- and two-family *dwellings* which are three stories or less in height and constructed with stud bearing walls, plain concrete footings without longitudinal reinforcement supporting walls and isolated plain concrete footings supporting columns or pedestals are permitted.

Discussion: In typical deck construction the most common foundation systems employed are footing foundations, pier foundations or a combination of both. The choice of which to use is usually related to the required depth of excavation, and the required bearing area based on total load and soil type. The IRC provides prescriptive design criteria for footing foundation systems, either directly supporting the deck post (column) at a point above or below grade or supporting a below grade pier. Regardless of the foundation design chosen, the bearing area in contact with the soil beneath the foundation is the only prescriptive means in the IRC to transfer loads to the soil. Deck "piers" are essentially just round or square concrete columns, often resting on a properly sized pad footing. Footings are generally designed to receive a load from a column or pier that bears on the top of the footing. Footing construction is ideal when large bearing areas are required, due to heavy loads or poor soil bearing capacity or when site conditions allow for shallow excavation depths. A minimum footing thickness of 6 inches (152 mm) and a minimum projection of 2 inches (51 mm) is required by the IRC for wall footings, such as for the house foundation wall resting on a continuous concrete wall footing. The footing projection beyond the foundation cannot exceed the thickness of the footing as the diagonal tension stress could cause cracking of the footing. In most cases the relationship between the area of the load placed on the footing and the required bearing area at the base of the footing will dictate the projection and thus the thickness necessary for the footing to disperse the concentrated load without cracking (see Example 3-8). Although footing projection is not specifically addressed for deck footings, when using a combination of a pier resting on a concrete footing pad, the 2 inches (51 mm) of minimum projection is a good rule to follow. The footing projection can act as an anchor below the soil to provide increased resistance to uplift and overturning.

When minimal bearing area or a deep frost depth is required, a monolithically poured footing, resembling a pier, may be more desirable. In these cases there is no need to calculate the relationship of the projection to the thickness, as the width of the bearing area at the base of the pier will be maintained throughout its height. If a large bearing area is required, this type of footing can gradually taper in as it nears the surface, as to leave less exposed concrete at the top of the footing above grade, and a more attractive finish. This is often referred to as a belled bottom pier (see Example 3-9).

In an exception to IRC Section R403.1.3, footings, as applicable to the subject matter of this book, do not require any reinforcing steel.

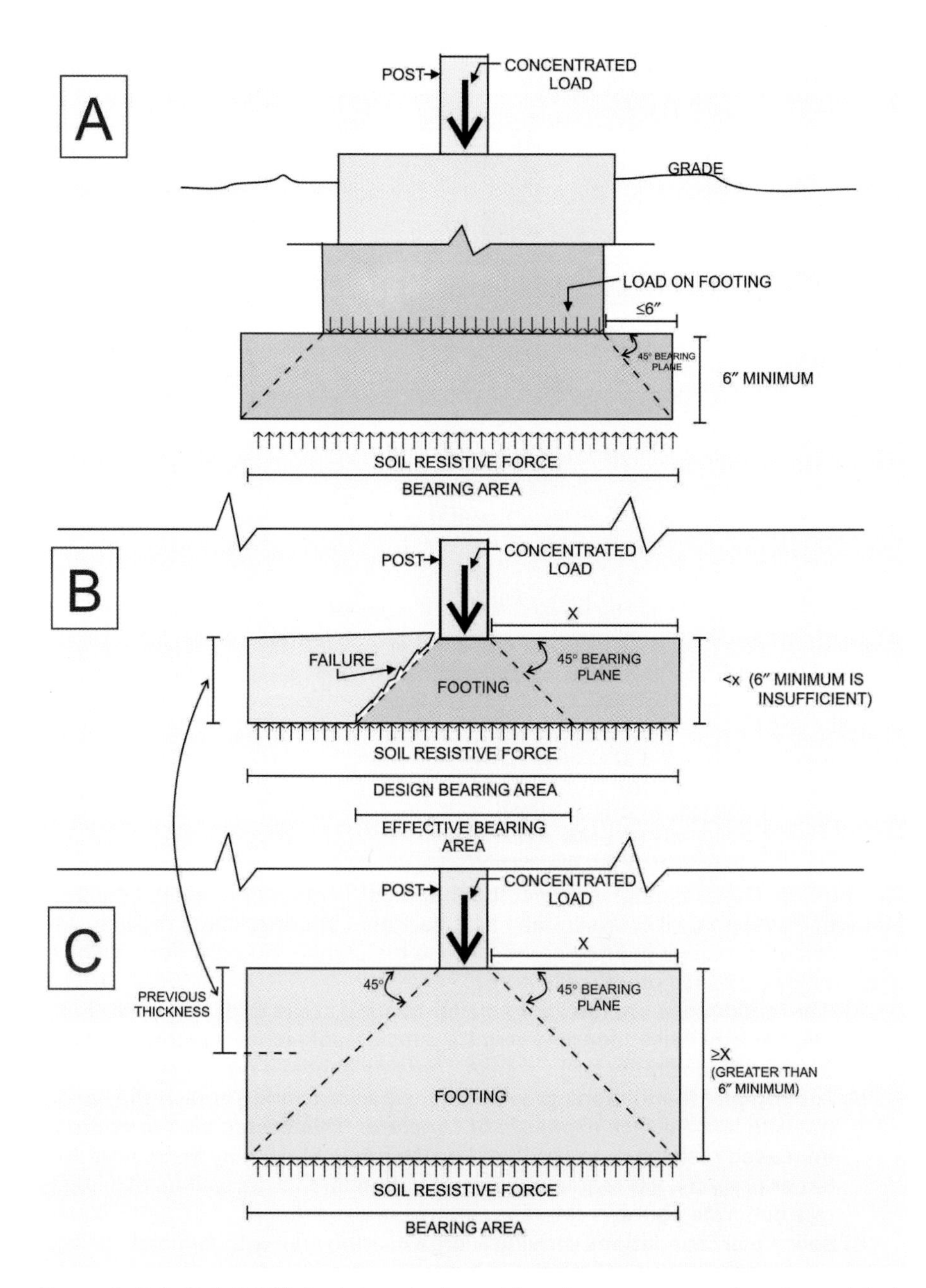

Example 3-8: Detail "A" depicts a deck post bearing on a concrete column (pier), supported by an enlarged footing. This design is convenient when a large bearing area is required. In detail "A" the 6-inch (152 mm) minimum thickness is acceptable. In detail "B" the concrete column has been omitted and the deck post is bearing directly on the top of the footing. In this arrangement the minimum 6-inch (152 mm) thickness is insufficient and may result in cracking of the footing along the bearing plane, thus reducing the effective bearing area of the footing. Detail "C" shows how the thickness of the footing was increased to at least equal with the projection in order to minimize the possibility for failure due to cracking.

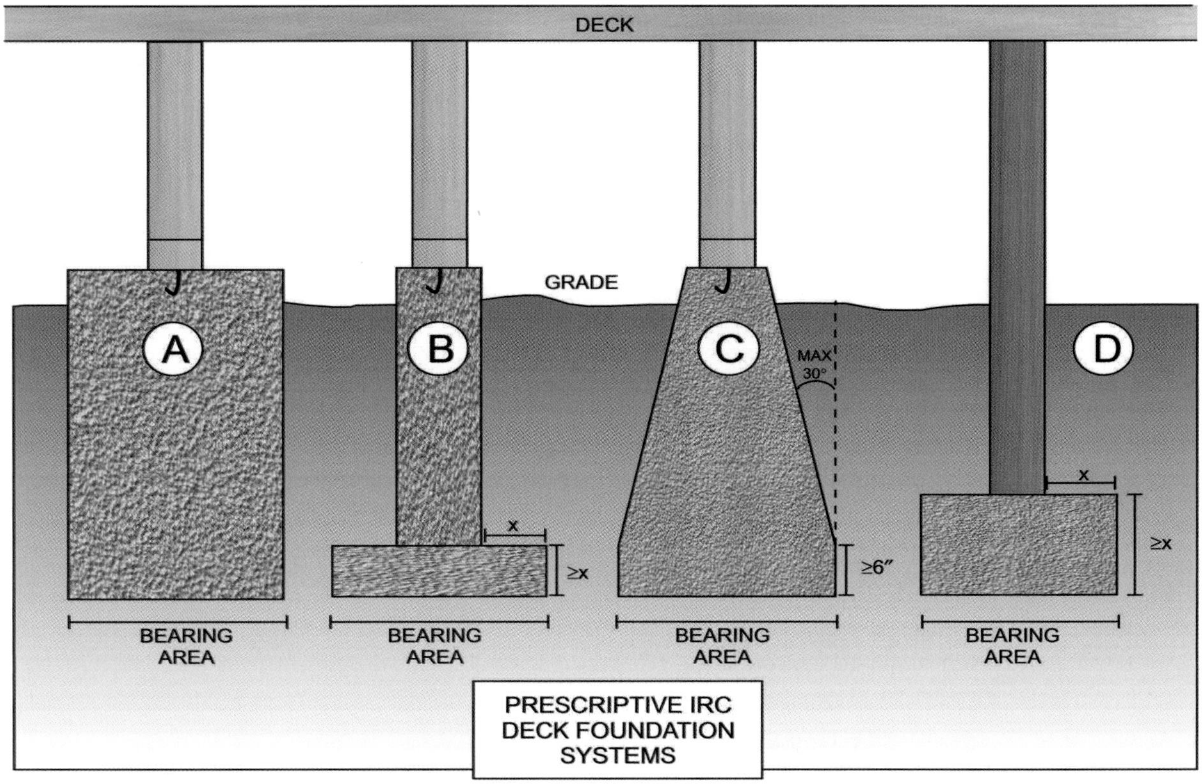

PRESCRIPTIVE IRC DECK FOUNDATION SYSTEMS

Example 3-9: These examples depict the four most common residential deck foundation systems, all of which may be acceptable options. All of these examples provide equivalent bearing areas for load resistance, yet selection of which to use may vary based on other features.

A) Pier foundations are ideal for smaller bearing areas or deep excavation depths; otherwise they may require considerably more concrete and leave an undesirably large area of concrete exposed on the surface.

B) Footing-pier foundations provide an increased bearing area at the base, but with less surface exposure of concrete. This design also provides increased resistance to uplift and overturning which may be desired for larger projects, yet requires overexcavation and fill, as well as two separate concrete pours.

C) Belled pier foundations provide a large bearing area with minimal surface exposure, similar to footing-pier foundations, but without overexcavation and fill. These foundations require care not to loosen the sidewalls, but can be completed in a single pour.

D) Footing foundations are convenient for shallow foundation excavations without a frost depth. When excavating to a frost depth, this foundation will require less concrete and may be a more affordable option. However, the post must be treated for ground contact use (see Chapter 4), and will be difficult to replace in the future.

Excerpt from the 2009 International Building Code Regarding Quality of Concrete Placement

1808.8.3 Placement of concrete. Concrete shall be placed in such a manner as to ensure the exclusion of any foreign matter and to secure a full-size foundation. Concrete shall not be placed through water unless a tremie or other method *approved* by the *building official* is used. Where placed under or in the presence of water, the concrete shall be deposited by *approved* means to ensure minimum segregation of the mix and negligible turbulence of the water. Where depositing concrete from the top of a deep foundation element, the concrete shall be chuted directly into smooth-sided pipes or tubes or placed in a rapid and continuous operation through a funnel hopper centered at the top of the element.

Discussion: The above excerpt is from the *International Building Code* (IBC) and provides guidance for proper placement of concrete in an earth excavation. Concrete is a mixture of water, cement, course aggregate (gravel) and sand. When placing concrete in a pier or footing excavation, the integrity of the concrete mix should be maintained. Pouring uncured concrete into water can cause the cement to dilute and separate from the aggregate. The previously homogenous mixture that makes concrete what it is will essentially become a bunch of rocks covered in cement, and will greatly reduce the strength and longevity of the pier.

Careless placement of the concrete mix for a pier that knocks soil, rocks, mulch or any other foreign debris into the blend of uncured concrete should be avoided. While typical residential deck construction won't realistically involve a funnel hopper, but rather a wheelbarrow or shovel, patience and care still must be employed. Obviously a "rapid and continuous" placement of concrete is infeasible with the premixed, bagged concrete typically used for deck construction, but proper preparations should be employed so that a close-to-continuous pour can still be achieved, without surface curing between wet concrete deposits.

Side Note: After a poorly timed thunderstorm an old plastic cup screwed to the end of a stick is a helpful tool to have around for removing water prior to a pour.

Part Three: Other Considerations

R401.3 Drainage. Surface drainage shall be diverted to a storm sewer conveyance or other *approved* point of collection that does not create a hazard. *Lots* shall be graded to drain surface water away from foundation walls. The *grade* shall fall a minimum of 6 inches (152 mm) within the first 10 feet (3048 mm).

> **Exception:** Where *lot lines*, walls, slopes or other physical barriers prohibit 6 inches (152 mm) of fall within 10 feet (3048 mm), drains or swales shall be constructed to ensure drainage away from the structure. Impervious surfaces within 10 feet (3048 mm) of the building foundation shall be sloped a minimum of 2 percent away from the building.

Discussion: Any construction project that involves excavating, moving or otherwise disturbing the soil surrounding a building foundation must be performed with consideration of this IRC section. To maintain the integrity and function of building foundation walls, the IRC requires that all grade adjacent to a foundation wall be sloped such that water will drain away. The final grade must fall at least 6 inches (152 mm) within 10 feet (3048 mm) from a foundation wall. This fall may occur in a steep slope occurring over just a few feet or a gradual slope over the entire 10 feet (3048 mm), but at the 10 foot (3048 mm) point, the grade must be at least 6 inches (152 mm) lower than at the foundation wall (see Example 3-10). Where this is not possible, an exception allows the use of draws or swales to ensure drainage away from the structure.

Often in deck construction, it is convenient to leave the soil from the foundation excavations underneath a low deck rather than removing it from the site. When this is done, the required minimum fall of the final grade must be maintained. Often, when older decks are removed for replacement, the backfilled soil against the foundation may have settled and left a depression. In these situations, it is recommended to leave the soil behind and attempt to re-create the fall in grade and fill in the depression.

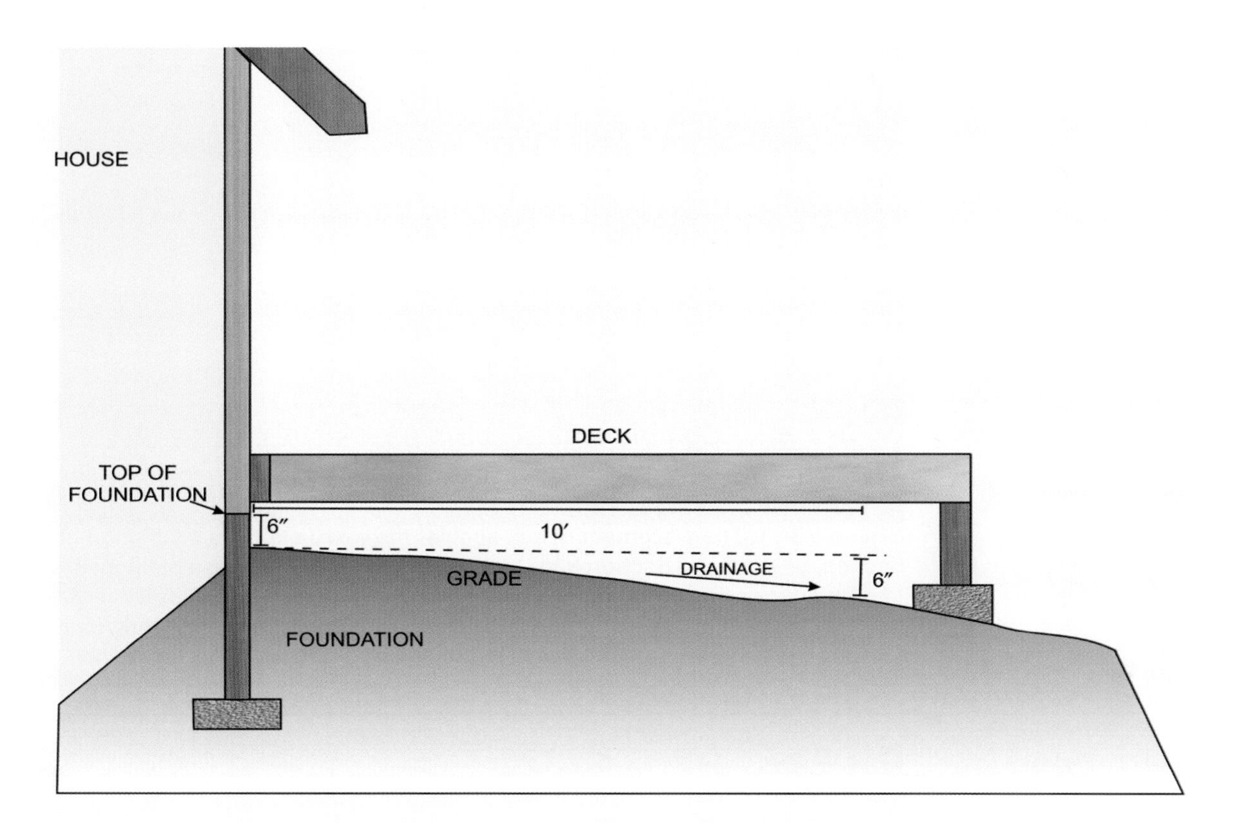

Example 3-10: To inhibit the absorption of liquid water near the foundation of the existing structure, and the increased pressure associated with it, the final grade must provide drainage away from the foundation.

R404.1.6 Height above finished grade. Concrete and masonry foundation walls shall extend above the finished *grade* adjacent to the foundation at all points a minimum of 4 inches (102 mm) where masonry veneer is used and a minimum of 6 inches (152 mm) elsewhere.

Discussion: If the excavated soil is left onsite against the foundation and the required fall discussed in the previous section can be maintained, there is still a limit to the height the soil can be placed. Foundation walls must extend at least 6 inches (152 mm) above the final adjacent grade for decay and termite protection of the exterior cladding (see Example 3-11). In the event that wood siding extends down past the top of the foundation the 6-inch (152 mm) clearance must be measured from the bottom of the siding in accordance with Section R318. In the case of adhered or anchored masonry veneer, the 6-inch (152 mm) clearance can be reduced to 4 inches (102 mm).

Example 3-11: The placement of soil against the wood siding is a violation of Section R404.1.6 and Section R318, and will dramatically increase the possibility of future decay occurring to the siding and floor framing beyond. If left in this manner, the deck would conceal the damage from the exterior, resulting in significant repair by the time the decay is noticed from the inside of the house.

Underground Utilities

Gas Pipe

G2415.10 (404.10) Minimum burial depth. Underground *piping systems* shall be installed a minimum depth of 12 inches (305 mm) below grade, except as provided for in Section G2415.10.1.

G2415.10.1 (404.10.1) Individual outside appliances. Individual lines to outside lights, grills or other *appliances* shall be installed a minimum of 8 inches (203 mm) below finished grade, provided that such installation is *approved* and is installed in locations not susceptible to physical damage.

Water Service Pipe

P2603.6 Freezing. In localities having a winter design temperature of 32°F (0°C) or lower as shown in Table R301.2(1) of this code, a water, soil or waste pipe shall not be installed outside of a building, in exterior walls, in *attics* or crawl spaces, or in any other place subjected to freezing temperature unless adequate provision is made to protect it from freezing by insulation or heat or both. Water service pipe shall be installed not less than 12 inches (305 mm) deep and not less than 6 inches (152 mm) below the frost line.

Sewer Pipe

P2603.6.1 Sewer depth. *Building sewers* that connect to private sewage disposal systems shall be a minimum of [NUMBER] inches (mm) below finished *grade* at the point of septic tank connection. *Building sewers* shall be a minimum of [NUMBER] inches (mm) below *grade.*

Electrical Cable

TABLE E3803.1
MINIMUM COVER REQUIREMENTS, BURIAL IN INCHES[a, b, c, d, e]

LOCATION OF WIRING METHOD OR CIRCUIT	TYPE OF WIRING METHOD OR CIRCUIT				
	1 Direct burial cables or conductors	2 Rigid metal conduit or intermediate metal conduit	3 Nonmetallic raceways listed for direct burial without concrete encasement or other approved raceways	4 Residential branch circuits rated 120 volts or less with GFCI protection and maximum overcurrent protection of 20 amperes	5 Circuits for control of irrigation and landscape lighting limited to not more than 30 volts and installed with type UF or in other identified cable or raceway
All locations not specified below	24	6	18	12	6
In trench below 2-inch-thick concrete or equivalent	18	6	12	6	6
Under a building	0 (In raceway only)	0	0	0 (In raceway only)	0 (In raceway only)
Under minimum of 4-inch- thick concrete exterior slab with no vehicular traffic and the slab extending not less than 6 inches beyond the underground installation	18	4	4	6 (Direct burial) 4 (In raceway)	6 (Direct burial) 4 (In raceway)
Under streets, highways, roads, alleys, driveways and parking lots	24	24	24	24	24
One- and two-family dwelling driveways and outdoor parking areas, and used only for dwelling-related purposes	18	18	18	12	18
In solid rock where covered by minimum of 2 inches concrete extending down to rock	2 (In raceway only)	2	2	2 (In raceway only)	2 (In raceway only)

For SI: 1 inch = 25.4 mm.
a. Raceways approved for burial only where encased concrete shall require concrete envelope not less than 2 inches thick.
b. Lesser depths shall be permitted where cables and conductors rise for terminations or splices or where access is otherwise required.
c. Where one of the wiring method types listed in columns 1 to 3 is combined with one of the circuit types in columns 4 and 5, the shallower depth of burial shall be permitted.
d. Where solid rock prevents compliance with the cover depths specified in this table, the wiring shall be installed in metal or nonmetallic raceway permitted for direct burial. The raceways shall be covered by a minimum of 2 inches of concrete extending down to the rock.
e. Cover is defined as the shortest distance in inches (millimeters) measured between a point on the top surface of any direct-buried conductor, cable, conduit or other raceway and the top surface of finished grade, concrete, or similar cover.

Discussion: The various IRC sections above provide an insight into the minimum allowable burial depths for services installed by other trade professionals. The life and property hazard of excavating a foundation system into one of these services should require no explanation; electricity and gas can cause pain and fire, and water and sewer make big ugly messes of the project site. Even if those calamities are averted, many of these services can only be repaired by state licensed electricians or plumbers, professionals who are not likely on a deck company's payroll. Other services may be owned and regulated by the utility provider, again requiring the service of additional professionals at an additional and unplanned expense.

A common burial depth of 12 inches (305 mm) is predominant among these services for the purpose of providing protection against physical damage. While this depth may protect the underground services from the shovel of a gardener, the pole of a volleyball net or the pounding of feet in the backyard, it is not deep enough to protect from the excavation of any foundation system. Even without a required frost depth, the minimum required depth of 12 inches (305 mm) into undisturbed soil, as discussed in Section R403.1.4, is enough to cause an unexpected and unpleasant encounter with these services (see Example 3-12).

Services that require protection from freezing, such as water and sewer pipe, are generally required to be at or below the frost depth. This depth is less likely to be struck in excavation, but should not be relied upon. For gas pipe serving an exterior appliance, such as a barbeque grill, fire pit or lantern, the 12-inch (305 mm) depth for protection against physical damage can be reduced to 8 inches (203 mm) below grade.

Example 3-12: The burial depth of this plastic natural gas pipe and electrical service conductors is a mere 27 inches (686 mm), and could have been as shallow as 12 inches (305 mm) below the final grade.

Side Note: In 2000 the Common Ground Alliance (CGA) formed and shortly after released a new nationwide underground locate service. A simple call to 811 anywhere in the US, including Alaska and Hawaii, will allow you to schedule an underground location marking of the local utilities. For more information, visit www.call811.com.

Chapter 4: Framing

Introduction

One of the most basic and fundamental expectations of the IRC, or for that matter any other model building code, is that of structural stability. No other IRC requirements are of any substantial importance if the structure to which they apply is not capable of staying in place. If a deck comes crashing down beneath your feet, the availability of guards, stairway illumination or safety glazing becomes considerably less important. If stairs collapse upon use, does the geometry of the tread depth and height really matter?

In recent years, deck instability and structural failure has been a popular topic in the media. Some of these failures were a result of poor building maintenance, as few products will last forever without care and attention, but many of them were due to an incomplete or insufficient load path. The term "load path" refers to the various structural members and connections that transfer loads from their point of origin to the resisting element, the earth.

All forces imposed on a structure must be transmitted through the structure to the supporting soil below. Whether the weight of a table on the deck, the force of someone leaning against a rail, or the uplifting pressure of the wind beneath an elevated deck, there must be a load path designed to transmit these forces to the supporting ground.

The basic organization of the structural framing components of a deck, such as joists, beams and posts, are not the only considerations of the load path. Individual material properties of hardware, hangers, lumber and connections must be considered, as well as the longevity of these materials. A deck built of inferior materials may be stable at the time of inspection, but how will it perform after a few years of exterior exposure to moisture, UV degradation or anything else that Mother Nature can subject it to?

One of the major failures of the load path in deck construction is the deck ledger to house connection, where components of the existing structure complete part of the deck's load path. However, these components are typically hidden behind exterior cladding such as brick or stucco or interior finishes such as drywall, leaving many unknowns in the load path. The connection of the ledger to the existing structure does not complete the load path, however. In many deck designs the existing structure may not be capable of resisting the new loads from the deck attachment. This is often the case in ledger connections to existing floor cantilevers or when significant lateral loads are applied to the existing band joist.

Even when the ledger connection is structurally adequate for all code-prescribed loading conditions and the appropriate materials have been used, there is still another potential issue to consider. When attaching a ledger to an existing structure, you are likely altering the exterior cladding of a structure, which is usually the first line of defense designed to resist the entry of liquid water into the structure's cavity. If water is able to penetrate the joint made between the new deck and the existing structure, even a properly designed connection can become prone to moisture damage and decay. Decay and weathering may deteriorate the materials to such a degree that they would not provide adequate support and may become prone to failure.

No model code or standards document, including the IRC, can be expected to provide comprehensive structural design criteria for every conceivable condition. When reviewing this chapter you will find a considerable amount of useful and applicable information in regard to structural design, yet you may also need clarification or provisions from additional sources for certain portions of the structure. Throughout this chapter and in the appendix, additional reputable sources will be suggested that may aid in the structural design of your deck. However, before moving forward using an alternative design document or absorbing the cost of engineering services, it is always prudent to consult your local building official as they may already have guidelines, policies or preapproved applications intended to aid you in using the IRC for your deck design.

Part One: Lumber Properties

Definitions

NATURALLY DURABLE WOOD. (IRC) The heartwood of the following species with the exception that an occasional piece with corner sapwood is permitted if 90 percent or more of the width of each side on which it occurs is heartwood.

Decay resistant. Redwood, cedar, black locust and black walnut.
Termite resistant. Alaska yellow cedar, redwood, Eastern red cedar and Western red cedar including all sapwood of Western red cedar.

TREATED WOOD. (IBC) Wood and wood-based materials that use vacuum-pressure impregnation processes to enhance fire retardant or preservative properties.

Fire-Retardant-Treated Wood. Pressure-treated lumber and plywood that exhibit reduced surface burning characteristics and resist propagation of fire.
Preservative-treated wood. Pressure-treated wood products that exhibit reduced susceptibility to damage by fungi, insects or marine borers.

R502.1 Identification. Load-bearing dimension lumber for joists, beams and girders shall be identified by a grade *mark* of a lumber grading or inspection agency that has been *approved* by an accreditation body that complies with DOC PS 20. In lieu of a grade *mark*, a certificate of inspection issued by a lumber grading or inspection agency meeting the requirements of this section shall be accepted.

R502.1.1 Preservative-treated lumber. Preservative treated dimension lumber shall also be identified as required by Section R319.1.

R502.1.2 Blocking and subflooring. Blocking shall be a minimum of utility grade lumber. Subflooring may be a minimum of utility grade lumber or No. 4 common grade boards.

R502.1.3 End-jointed lumber. *Approved* end-jointed lumber identified by a grade *mark* conforming to Section R502.1 may be used interchangeably with solid-sawn members of the same species and grade.

R502.1.4 Prefabricated wood I-joists. Structural capacities and design provisions for prefabricated wood I-joists shall be established and monitored in accordance with ASTM D 5055.

R502.1.5 Structural glued laminated timbers. Glued laminated timbers shall be manufactured and identified as required in ANSI/AITC A190.1 and ASTM D 3737.

R502.1.6 Structural log members. Stress grading of structural log members of nonrectangular shape, as typically used in log buildings, shall be in accordance with ASTM D 3957. Such structural log members shall be identified by the grade *mark* of an *approved* lumber grading or inspection agency. In lieu of a grade *mark* on the material, a certificate of inspection as to species and grade issued by a lumber-grading or inspection agency meeting the requirements of this section shall be permitted to be accepted.

Discussion: Section R502.1 and all of its subsections detail the grade, test standards and identification required for the various types of lumber products. All dimensional lumber used for structural applications must have a grade mark from an accredited lumber grading or inspection agency which identifies various properties of the lumber, such as the wood species, grade and moisture content. Pressure preservative treatments must comply with these same marking requirements, as well as additional identification marks by an approved inspection agency that will be discussed in the following section. End-jointed lumber, also not common in deck construction, may also be used in various applications in accordance with the information provided in the grade mark. The species, grade and moisture content of the structural material is vital to proper use of pre-engineered span tables or other design documents, as they affect the structural capacities of the material. This will be evident later in this chapter when reviewing the joist span tables provided in the IRC.

Engineered and manufactured wood products such as I-joists and glued-laminated beams may also be used in deck construction, provided they comply with the respective standards listed above. Generally, I-joists are not approved for use in wet locations, and would only be effective in applications with water-tight decking surfaces, and concealed from below. As discussed further in this chapter, the properties, structural capacities and allowable use of I-joists and other engineered lumber products can vary greatly from dimensional lumber. Glued-laminated timbers require marking in accordance with the above mentioned standards to ensure proper installation in the field (see Example 4-1). The structural properties of glued-laminated beams depend on the species and grade of the lumber used in the laminations, which is reflected by the combination symbol. This will sometimes result in a glued-laminated beam being marked with one face that must be installed at the top to ensure that the tension lamination is on the bottom of the beam. Natural log timbers must also undergo a grading inspection so that their properties can be verified through a grade marking.

Example 4-1: The markings visible in this photo of a glued-laminated beam only provide a portion of the required information.

R317.1 Location required. Protection of wood and wood based products from decay shall be provided in the following locations by the use of naturally durable wood or wood that is preservative-treated in accordance with AWPA U1 for the species, product, preservative and end use. Preservatives shall be listed in Section 4 of AWPA U1.

4. The ends of wood girders entering exterior masonry or concrete walls having clearances of less than $^1/_2$ inch (12.7 mm) on tops, sides and ends.
6. Wood structural members supporting moisture-permeable floors or roofs that are exposed to the weather, such as concrete or masonry slabs, unless separated from such floors or roofs by an impervious moisture barrier.

R317.1.2 Ground contact. All wood in contact with the ground, embedded in concrete in direct contact with the ground or embedded in concrete exposed to the weather that supports permanent structures intended for human occupancy shall be *approved* pressure-preservative-treated wood suitable for ground contact use, except untreated wood may be used where entirely below groundwater level or continuously submerged in fresh water.

R317.1.3 Geographical areas. In geographical areas where experience has demonstrated a specific need, *approved* naturally durable or pressure-preservative-treated wood shall be used for those portions of wood members that form the structural supports of buildings, balconies, porches or similar permanent building appurtenances when those members are exposed to the weather without adequate protection from a roof, eave, overhang or other covering that would prevent moisture or water accumulation on the surface or at joints between members. Depending on local experience, such members may include:

1. Horizontal members such as girders, joists and decking.
2. Vertical members such as posts, poles and columns.
3. Both horizontal and vertical members.

R317.1.4 Wood columns. Wood columns shall be *approved* wood of natural decay resistance or *approved* pressure-preservative-treated wood.

Exceptions:

1. Columns exposed to the weather or in *basements* when supported by concrete piers or metal pedestals projecting 1 inch (25.4 mm) above a concrete floor or 6 inches (152 mm) above exposed earth and the earth is covered by an *approved* impervious moisture barrier.
2. Columns in enclosed crawl spaces or unexcavated areas located within the periphery of the building when supported by a concrete pier or metal pedestal at a height more than 8 inches (203 mm) from exposed earth and the earth is covered by an impervious moisture barrier.

R317.1.5 Exposed glued-laminated timbers. The portions of glued-laminated timbers that form the structural supports of a building or other structure and are exposed to weather and not properly protected by a roof, eave or similar covering shall be pressure treated with preservative, or be manufactured from naturally durable or preservative-treated wood.

Discussion: The five IRC sections provided above all serve to detail the different situations in which lumber, when used, must be decay-resistant to provide a long service life in a moist or otherwise harsh environment. Section R317.1 provides the IRC reference to the American Wood Protection Association's standard U1 as the basis for approving the use of different wood species preserved using various treatments and treatment methods, as described within the U1 standard. Many of the common treated wood products on the market are contained within the U1 standard. The IRC sections following Section R317.1 use the adjective "approved" when describing the need for preservative-treated material, and this approval will likely be based on the AWPA U1 standard as well. Other than for ground contact conditions, as described in Section R317.1.2 and explained later in this discussion, naturally durable wood can be used in other locations where decay resistance is needed. Naturally durable wood is specifically defined by the IRC to include only limited species of wood when 90 percent or more heartwood is present on each side of the member (see definitions).

In Section R317.1, Item 4 requires decay-resistant wood when the ends enter into exterior masonry or concrete walls, such as the deck joists in Example 4-2. Item 6 requires a decay-resistant structure when a slab-type moisture permeable deck surface is installed, such as concrete, masonry or tile. This less-common deck installation may result in moisture being trapped between the decking surface and the supporting panels below, thus accelerating the possibility of moisture-related damage. However, if a moisture-impervious barrier is installed to separate the decking material from its supporting structure, the possibility for decay is diminished, and standard wood material may be approved for use. There are many products available as a moisture-impervious barrier, both solid and liquid applied. While these two listed conditions apply directly to exterior deck construction, many of the conditions listed under Section R317.1 are precluded by the remaining IRC sections provided above.

Glued-laminated lumber, as described in Section R317.1.5, must be preservative-treated or constructed of naturally durable lumber when exposed to the weather.

Section R317.1.4 refers specifically to wood support columns. Unlike an entire wall, with multiple members (studs) providing support at intervals of no more than 24 inches (610 mm), a post is a stand-alone structural member. Significant decay at the base or any other portion of a wood column has the potential to create a catastrophic failure due to the lack of any other nearby supporting member. The IRC requires wood columns to be decay-resistant material. The exceptions to this section do not apply to deck construction, as they require the use of a moisture barrier over the ground adjacent to the post...not something practical for deck construction.

Unlike any of the other related IRC sections, Section R317.1.3 is completely based on local experience and each individual building department's determination. This section may preclude all the others, as all of the wood structural components that make up a deck structure may be required to be naturally durable lumber or pressure-preservative-treated lumber. Many times throughout the code, local climatic or geographic criteria must be determined prior to using the related code provisions. Table R301.2(1), provided in Part 3 of this chapter, provides most of that criteria.

Section R317.1.2 describes conditions when wood is in contact with the earth, or embedded in concrete in contact with the earth. These conditions are the only ones where naturally durable wood is not acceptable as a means of decay resistance. Concrete is porous and the ground is slow to dry. Combining these characteristics places lumber in contact with moisture for periods of time in excess of the decay-resistant capabilities of naturally durable lumber (see Example 4-3).

Not all lumber or treatments perform the same. In order to be labeled for ground contact conditions, the treatment retention levels must be appropriate for the lumber species and the type of treatment chemical used, as described within the AWPA U1. Wood placed in ground-contact conditions must be verified through ground-contact identification on the grading label.

Although good practice for any lumber in ground-contact conditions, this requirement specifically addresses "permanent structures intended for human occupancy." Whether that would include shading trellises or arbors, fences, benches or other such amenities built into the earth, would be left to the designer of the project.

Example 4-2: This deck was constructed at the same time as the house, allowing the joists to be connected directly to the band joist. The brick veneer was then installed and cut around the deck joists. Wood joists in this condition must be pressure-preservative treated or naturally durable.

Example 4-3: This untreated Douglas fir post, embedded in concrete in contact with the earth does not comply with Section R319.1.2, as the post must be pressure-preservative-treated lumber.

R318.1.1 Quality mark. Lumber and plywood required to be pressure-preservative-treated in accordance with Section R318.1 shall bear the quality *mark* of an *approved* inspection agency that maintains continuing supervision, testing and inspection over the quality of the product and that has been *approved* by an accreditation body which complies with the requirements of the American Lumber Standard Committee treated wood program.

R317.2.1 Required information. The required quality *mark* on each piece of pressure-preservative-treated lumber or plywood shall contain the following information:

1. Identification of the treating plant.
2. Type of preservative.
3. The minimum preservative retention.
4. End use for which the product was treated.
5. Standard to which the product was treated.
6. Identity of the *approved* inspection agency.
7. The designation "Dry," if applicable.

Exception: Quality *marks* on lumber less than 1 inch (25.4 mm) nominal thickness, or lumber less than nominal 1 inch by 5 inches (25.4 mm by 127 mm) or 2 inches by 4 inches (51 mm by 102 mm) or lumber 36 inches (914 mm) or less in length shall be applied by stamping the faces of exterior pieces or by end labeling not less than 25 percent of the pieces of a bundled unit.

Discussion: The previously discussed sections provide guidance for when preservative-treated lumber must be used (ask the building official, or always use it), but how can its use be verified? The days of obvious green lumber to prove the presence of pressure-treatment are slowly coming to an end. Since the voluntary removal of CCA-treated material from the residential market, many new treatment methods are available and the market share has been dispersed among them. It is sometimes difficult to recognize some treated material from untreated material, particularly when some of the chemical preservatives may appear in green and brown shades and some are almost clear treatments. All treated lumber used in deck construction must be bear the mark of an approved inspection agency. This mark is required to verify that the material has been appropriately treated, what treatment was used, the level of treatment retention, and most importantly, the allowable use of the product, such as for ground contact or above ground use. This marking may come in the form of a stamp on the lumber or a tag stapled to one end (see Example 4-4).

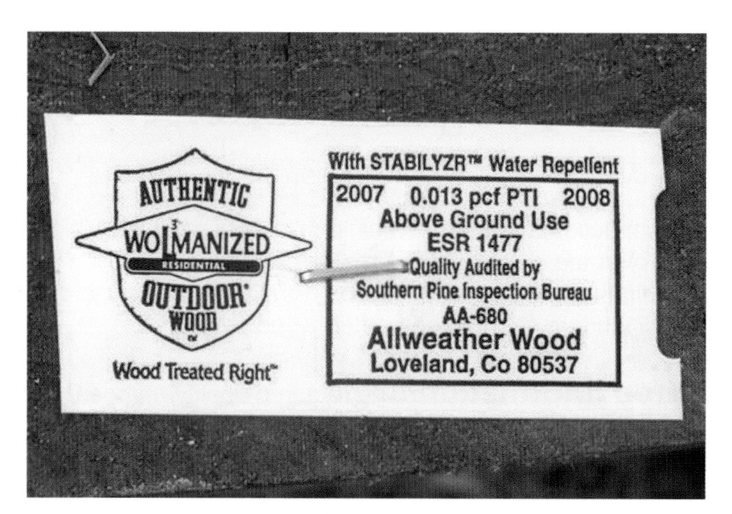

Example 4-4: Some treated lumber will have an identifying tag stapled to one end. This tag provides the necessary information for approval and proper use of the lumber. The cut-off ends of treated lumber with this type of tag should be kept onsite until verified by the inspector.

R317.1.1 Field treatment. Field-cut ends, notches and drilled holes of preservative-treated wood shall be treated in the field in accordance with AWPA M4.

Discussion: Unlike the heartwood of naturally durable lumber, preservative-treated lumber is not homogenously decay resistant. If you've cut certain treated lumber before, you've probably seen the green ring of treatment around a white core of untreated lumber in the cross-section of that cut. Preservative treatment in lumber does not often reach the middle of the cross section. The depth of penetration varies depending on the wood species, cross-sectional dimensions and the proposed end-use noted on the tag. When drilling, cutting, notching or otherwise altering treated material in any way that exposes previously internal parts, those parts must be treated in the field.

The American Wood Protection Association (AWPA) is the recognized authority of wood preservative treatment in the United States, and publishes many standards in regard to the subject. M4, *Care of Preservative-Treated Wood Products*, is a standard published by the AWPA, and is referenced for the field treatment required by this section. While this standard specifies good practices for properly storing treated lumber on the jobsite, it is only referenced by the IRC for the purpose of field treatment.

AWPA M4 reinforces the requirement of the IRC to field treat all internal and untreated areas that may be exposed by alterations to the lumber. The treatments must be applied in accordance with the manufacturer's requirements, and must be applied before installation. However, what is most important is that it is applied with care, and all excess, unabsorbed treatment is wiped off the material. Decks with cantilevers over dropped beams are often easier to build if the joists are installed long, and then trimmed off of the cantilevered end. In this case, treatment would simply need to be applied to the cut ends prior to installing the rim joist.

In the case of bored or drilled holes for the installation of connectors, such as lag bolts at the ledger, or bolts at the guard posts, there is an alternative to the treatments listed in the standard. In these areas, coal-tar roofing cement can be injected into the hole prior to installing the bolt.

Of the three field treatment materials provided in this standard, only one treatment is readily available and applicable to the common preservative-treated materials typically found in deck construction—copper naphthenate. Copper naphthenate can be found at most large home-improvement stores. Read and employ the product warnings carefully, including the use of protective clothing and maintaining control of the substance.

R318.1 Subterranean termite control methods. In areas subject to damage from termites as indicated by Table R301.2(1), methods of protection shall be one of the following methods or a combination of these methods:

1. Chemical termiticide treatment, as provided in Section R318.2.
2. Termite baiting system installed and maintained according to the *label*.
3. Pressure-preservative-treated wood in accordance with the provisions of Section R317.1.
4. Naturally durable termite-resistant wood.
5. Physical barriers as provided in Section R318.3 and used in locations as specified in Section R318.1.
6. Cold-formed steel framing in accordance with Sections R505.2.1 and R603.2.1.

R318.1.1 Quality mark. Lumber and plywood required to be pressure-preservative-treated in accordance with Section R318.1 shall bear the quality *mark* of an *approved* inspection agency which maintains continuing supervision, testing and inspection over the quality of the product and which has been *approved* by an accreditation body which complies with the requirements of the American Lumber Standard Committee treated wood program.

R318.1.2 Field treatment. Field-cut ends, notches, and drilled holes of pressure-preservative-treated wood shall be retreated in the field in accordance with AWPA M4.

R318.2 Chemical termiticide treatment. Chemical termiticide treatment shall include soil treatment and/or field applied wood treatment. The concentration, rate of application and method of treatment of the chemical termiticide shall be in strict accordance with the termiticide *label*.

R318.3 Barriers. Approved physical barriers, such as metal or plastic sheeting or collars specifically designed for termite prevention, shall be installed in a manner to prevent termites from entering the structure. Shields placed on top of an exterior foundation wall are permitted to be used only if in combination with another method of protection.

R318.4 Foam plastic protection. In areas where the probability of termite infestation is "very heavy" as indicated in Figure R301.2(6), extruded and expanded polystyrene, polyisocyanurate and other foam plastics shall not be installed on the exterior face or under interior or exterior foundation walls or slab foundations located below *grade*. The clearance between foam plastics installed above *grade* and exposed earth shall be at least 6 inches (152 mm).

Exceptions:

1. Buildings where the structural members of walls, floors, ceilings and roofs are entirely of noncombustible materials or pressure-preservative-treated wood.
2. When in *addition* to the requirements of Section R318.1, an *approved* method of protecting the foam plastic and structure from subterranean termite damage is used.
3. On the interior side of *basement walls*.

Discussion: Decks may undergo premature deterioration if unprotected and subject to termite infestation. Table R301.2(1), provided and discussed later in this chapter, contains provisions that vary based on the geographical location and historical evidence of a specific jurisdiction. The information in this table must be provided by the local jurisdiction, and will include information regarding the prevalence of termite damage to structures and the need to protect wood materials.

The most common and feasible means to protect a deck structure from termite damage is the use of pressure-preservative-treated lumber or lumber with natural resistance to termites, as detailed in the IRC definition of "Naturally durable wood." The other options provided by the IRC in this section are not as commonly employed for exterior deck applications and are outside the scope of this book.

Part Two: Fasteners and Hardware

R317.3 Fasteners and connectors in contact with preservative-treated and fire-retardant-treated wood.

Fasteners and connectors in contact with preservative-treated wood and fire-retardant-treated wood shall be in accordance with this section. The coating weights for zinc-coated fasteners shall be in accordance with ASTM A 153.

R317.3.1 Fasteners for preservative-treated wood.

Fasteners for preservative-treated wood shall be of hot dipped zinc-coated galvanized steel, stainless steel, silicon bronze or copper. Coating types and weights for connectors in contact with preservative-treated wood shall be in accordance with the connector manufacturer's recommendations. In the absence of manufacturer's recommendations, a minimum of ASTM A 653 type G185 zinc-coated galvanized steel, or equivalent, shall be used.

Exceptions:

1. One-half-inch (12.7 mm) diameter or greater steel bolts.
2. Fasteners other than nails and timber rivets shall be permitted to be of mechanically deposited zinc coated steel with coating weights in accordance with ASTM B 695, Class 55 minimum.

Discussion: Many wood treatments use copper as the primary fungicide and insecticide. The difficulty with using this component in treated lumber is that copper and steel don't mix.

Corrosion can occur when two dissimilar metals are in contact with one another. A small electric current flows from one metal to the other, accelerating corrosion of the more reactive metal. Of the metal selection available in this section for fasteners, stainless steel, silicon bronze and copper will experience very little to no oxidation when in contact with the copper in the preservative treatment, as they are very stable metals and alloys with low reactivity. Hot-dipped galvanized steel will experience oxidation, which may appear to be corroding and weakening the fastener. However, this is exactly what it is designed to do. The zinc coating from the galvanizing process is sacrificial, and after oxidizing it will create a protective layer of zinc-carbonate over the steel. If the weight of the zinc coating is not sufficient, the protective layer will not form to a sufficient thickness to protect the steel. ASTM A 153 is the standard for galvanized fasteners for use with preservative-treated material.

ASTM A 153 is a manufacturing standard and provides criteria that a manufacturer must satisfy in order to label their material in accordance with ASTM A 153. Builders and inspectors should verify whether the packaging for a zinc-coated fastener in contact with pressure-preservative-treated lumber is labeled as being tested in accordance with this standard.

It is important to realize that even without copper in the preservative treatment, there are other chemicals that may also be corrosive to highly reactive metals, such as steel. This section regulates fasteners used in any type of treated lumber, either for decay, termite or fire resistance, regardless of the chemical make-up of the treatment.

Two exceptions are provided to this requirement; the first one is applicable only to steel bolts $^{1}/_{2}$ inch (12.7 mm) diameter or greater. Just like the sacrificial zinc layer, the outermost steel in the fastener will oxidize first and will create a protective layer over the internal steel. It is recognized that the structural capacity of large diameter bolts will not be significantly diminished by the oxidation of the outer portions of the steel.

Although this exception applies to sill anchor bolts in a foundation wall protected from the weather, this exception does not apply to any diameter anchor bolts at an exterior ledger to band joist connection. Section R502.2.2.1, discussed in Part 4 of this chapter, has specific requirements for ledger fasteners that precludes this section. Remember, when two code sections contradict each other, the more specific section applies over the less specific.

The second exception allows the use of mechanically deposited zinc coatings in accordance with a different ASTM standard, yet limits the type of fasteners to those that would not typically be used in deck construction.

Side Note: Fastener manufacturers commonly make statements on their packaging such as, "approved for use with ACQ." This statement is misleading, as it implies that a product actually meets the required ASTM A 153 standard, when in fact it may not. Statements like this from manufacturers will likely lead to noncompliance of code provisions unless the user of the product verifies compliance with the standard.

Hardware and Framing Anchors

Section R317.3.1 requires the coatings on connectors, such as joist hangers and post connectors, to be in accordance with the manufacturer's recommendations for corrosion resistance when in contact with preservative-treated wood. Most major hardware manufacturers provide corrosion resistant recommendations based on the moisture level of the environment they are installed in and the specific chemical treatment of the lumber they are installed against. When a manufacturer has not provided recommendations applicable to a given installation, the connectors must be galvanized in accordance with ASTM A 653 type G185. Heavy galvanization on steel connectors provides a layer of zinc protection of the steel through the same chemical fundamentals explained earlier in this discussion.

For evaluating the structural performance, there are no specific criteria provided in the IRC for approving the use of hardware, such as joist hangers, post bases and similar components as the means to satisfy the load path and connection requirements discussed further in this chapter. The use of these products must be approved by the local building official and evidence for this approval must be provided just as for an "alternative" material or design (see Chapter 1).

The *International Building Code* (IBC), Chapter 17, provides code sections and references that can be used as guidance for an IRC approval of joist hanger use. These IBC sections refer to ASTM D 1761 as the criteria for evaluating a joist hanger in regard to resisting vertical loads and rotation. For manufactured products, an evaluation report issued by the ICC Evaluation Service (ICC-ES) or other approved independent third-party testing of a product is often obtained by product manufacturers for hardware and framing anchors such that allowable loads and installation methods can be determined. An agency, such as ICC Evaluation Service, as discussed further in Chapter 5, can be used to evaluate a product in accordance with the codes and referenced standards. This evaluation and/or testing will yield an evaluation report that a designer, installer and building official can reference for verifying the appropriate uses. The load tables and installation requirements published by hardware manufacturers are also a basis of the approval and must also be the basis for the installation (see Example 4-5). Custom fabricated hardware and brackets, not mass fabricated and distributed, are not as convenient to test for evidence of the structural capacities, due to the limited number produced. These products can be approved for use through the building official's review of an engineered design containing the seal of a registered design professional (see Example 4-6). A design professional, either private or employed by the manufacturer of the product, may also provide data to substantiate an alternative use of a manufactured product (see Example 4-7).

Example 4-5: The overdriven fasteners in this hanger may reduce the structural capacities of the steel due to deformation of the plane of steel against the ledger. Manufactured and tested hardware and framing anchors cannot be modified in any way unless specifically allowed by the manufacturer. Other installation limitations include the type of fasteners, size of lumber and contact of lumber to lumber in the connection. This information can usually be found in the general notes at the front of the major hardware manufacturer's product catalogs.

Example 4-6: A very unusual connection was made on this post that connects numerous construction elements together. As atypical as it is, a registered design professional determined it was sufficient for this specific application. The IRC simply seeks satisfactory structural performance of structures, regardless of how it is achieved.

Example 4-7: The post base in this photo was not installed in accordance with the manufacturer's installation instructions as more than half of its base was not bearing on the foundation below. A design professional provided documentation of this alternative method of installation as evidence that it will satisfy the site-specific loading requirements.

Part Three: Design

Definitions

Balloon framing. (McGraw-Hill) Framing for a building in which each stud is one piece from roof to foundation.

DEAD LOADS. (IRC) The weight of all materials of construction incorporated into the building, including but not limited to walls, floors, roofs, ceilings, stairways, built-in partitions, finishes, cladding, and other similarly incorporated architectural and structural items, and fixed service *equipment.*

LIGHT-FRAME CONSTRUCTION. (IRC) A type of construction whose vertical and horizontal structural elements are primarily formed by a system of repetitive wood or light gage steel framing members.

LIVE LOADS. (IRC) Those loads produced by the use and occupancy of the building or other structure and do not include construction or environmental loads such as wind load, snow load, rain load, earthquake load, flood load or dead load.

LOADS. (IBC) Forces or other actions that result from the weight of building materials, occupants and their possessions, environmental effects, differential movement and restrained dimensional changes. Permanent loads are those loads in which variations over time are rare or of small magnitude, such as dead loads. All other loads are variable loads (see also "*Nominal loads*").

PLATFORM CONSTRUCTION. (IRC) A method of construction by which floor framing bears on load bearing walls that are not continuous through the *story* levels or floor framing.

Post-and-beam construction. (McGraw-Hill) A type of wall construction using posts instead of studs.

POST-FRAME CONSTRUCTION. See "Post-and-beam construction."

REGISTERED DESIGN PROFESSIONAL. (IRC) An individual who is registered or licensed to practice their respective design profession as defined by the statutory requirements of the professional registration laws of the state or *jurisdiction* in which the project is to be constructed.

STRUCTURE. (IRC) That which is built or constructed.

VENEER. (IBC) A facing attached to a wall for the purpose of providing ornamentation, protection or insulation, but not counted as adding strength to the wall.

R301.1 Application. Buildings and structures, and all parts thereof, shall be constructed to safely support all loads, including dead loads, live loads, roof loads, flood loads, snow loads, wind loads and seismic loads as prescribed by this code. The construction of buildings and structures in accordance with the provisions of this code shall result in a system that provides a complete load path that meets all requirements for the transfer of all loads from their point of origin through the load-resisting elements to the foundation. Buildings and structures constructed as prescribed by this code are deemed to comply with the requirements of this section.

Discussion: A very general, yet powerful, provision, this section sets the foundation (pun intended) for the requirement that code-prescribed loads imposed on a structure be safely transmitted to the foundation and supporting soil. No matter the design, material, method of construction, construction element or location, the load path must be capable of being followed from any point of origin where loads may be imposed, through any other elements, including fasteners, hardware or other structural members, until the supporting foundation has been reached. Generally the framing system of the deck is the primary focus in terms of load path, but this section would also cover loads imposed on guards, handrails, stairs, benches, shading structures, fascia or anything else that may have loads imposed upon it.

The various code-prescribed loads are specifically mentioned in this IRC section, and when a structure is constructed under the prescriptive design criteria contained in this section, it is considered as compliant with the code. However, as presented throughout this book, no model code could feasibly provide comprehensive design criteria for all the various configurations and products used in deck construction. When employing other "alternative" designs, as described in Chapter 1 of this book, the load-path requirement of this section must be the cornerstone of the design.

R301.1.1 Alternative provisions. As an alternative to the requirements in Section R301.1 the following standards are permitted subject to the limitations of this code and the limitations therein. Where engineered design is used in conjunction with these standards the design shall comply with the *International Building Code.*

1. American Forest and Paper Association (AF&PA) *Wood Frame Construction Manual* (WFCM).
2. *American Iron and Steel Institute (AISI) Standard for Cold-Formed Steel Framing—Prescriptive Method for One- and Two-Family Dwellings* (AISI S230)
3. ICC-400 *Standard on the Design and Construction of Log Structures.*

Discussion: Do not let the use of the term "alternative" in this section be mistaken for "alternative materials, design and methods of construction and equipment," as described in Chapter 1 of this book. The prescriptive provisions contained in these three referenced documents are, in essence, part of the IRC.

The *Wood Frame Construction Manual* (WFCM) is published by the American Forest and Paper Association and is permitted as an alternative to the wood-frame construction provisions of the IRC. The WFCM does contain some additional deck-related provisions not provided in the IRC which will be discussed in the respective parts of this chapter.

In the event a deck is to be constructed using cold-formed steel, the American Iron and Steel Institute standard may be used in lieu of the prescriptive steel framing criteria provided in the IRC. Cold-formed steel frame design and log structure design, while provided in the IRC and the references above, are not very common in deck construction, and are thus outside the scope of this book.

R301.1.2 Construction systems. The requirements of this code are based on platform and balloon-frame construction for light-frame buildings. The requirements for concrete and masonry buildings are based on a balloon framing system. Other framing systems must have equivalent detailing to ensure force transfer, continuity and compatible deformations.

Discussion: A subsection to Section R301.1, this section describes the types of construction systems prescribed by the IRC for the purpose of fulfilling the load-path requirement established in the parent section (see the discussion of Section R301.1). Here is where typical deck construction separates itself from some of the prescriptive criteria in the IRC. The structural design provisions of the IRC are intended for use in platform and balloon framing—both of which refer to walls or studs within their definitions. It's rare that you will find or build an exterior deck that is supported by continuous wall sections on all sides. Deck construction is generally all post-frame construction (post-and-beam).

For vertical loads, imposed by gravity, there is not a tremendous difference in applying IRC provisions for post-frame construction to that of platform- or balloon-frame construction. Beams spanning between and over vertical posts will support live, dead, snow and roof loads in a similar manner to that of a window header in a bearing wall system. However, there are no engineering tables in the IRC for beam spans based on the design loads. The IRC table for solid-sawn wood wall header spans is based on total building width and the number of floors and the roof it supports; this information is not easily translated into applicable criteria for exterior deck beams. While other sources are necessary to calculate the beam span (see appendix), many of the remaining IRC provisions can still be applied for vertical loads.

Span tables for joists are also contained in the IRC, but likewise do not "fit" perfectly for exterior deck construction. Details of this are discussed under Section R502.3.2 in Part 5 of this chapter. Luckily, there are other sources of information on exterior beam and joist spans that can be submitted to the building official for consideration as an "alternative." The American Wood Council (the parent organization of the American Forest & Paper Association), the Western Wood Products Association and the Southern Pine Council are just a few reputable organizations that provide span tables appropriate for deck construction. These organizations are listed in the appendix of this book. Many lumber retailers and manufacturers may also provide span tables or computer-generated engineering for the lumber products they sell.

Generally, span tables are applicable only for uniformly distributed loads, based on different common joist spacing—12 inch (305 mm) on center, 16 inch (406 mm) on center and 24 inch (610 mm) on center, and the span of the joists and/or beams. If a deck is designed with any concentrated loads imposed within a beam span, such as posts for decks above or joists carrying other joists, then these tables may not apply, and specific engineering could be required for the loading conditions (see Example 4-8). There are some limited provisions in the IRC for carrying joists from other joists, such as framing around openings (see Section R502.10 in Part 4 of this chapter).

While many provisions in the IRC in regard to vertical loads can still apply to post-frame construction, lateral loads, such as those directed horizontally toward the structure, cannot. The lateral loads imposed on decks from wind and seismic forces vary based on the geographical and climatic conditions of the deck's location. Generally, these loads are minimal, and are only a major concern in particular designs and certain regions of the country. The movement of people on deck surfaces can also create lateral forces that can cause drift or sway in the deck surface. This movement typically increases as a deck ages and the wood shrinks and fasteners loosen. Simple methods can be used to resist these forces and prohibit the deck surface from swaying (see Examples 4-9, 4-10 and 4-11). Some methods of resisting lateral forces can be found in documents published by reputable organizations that can be approved by a local jurisdiction as an "alternative," such as the American Wood Council's DC A6 (see appendix).

Example 4-8: The roof above this deck was designed with a post bearing at the middle of the span of a deck beam. This beam cannot be sized using a beam span table due to this concentrated load. Designing posts that stack on one another and avoiding conditions like this can help avoid the need for job-specific engineering.

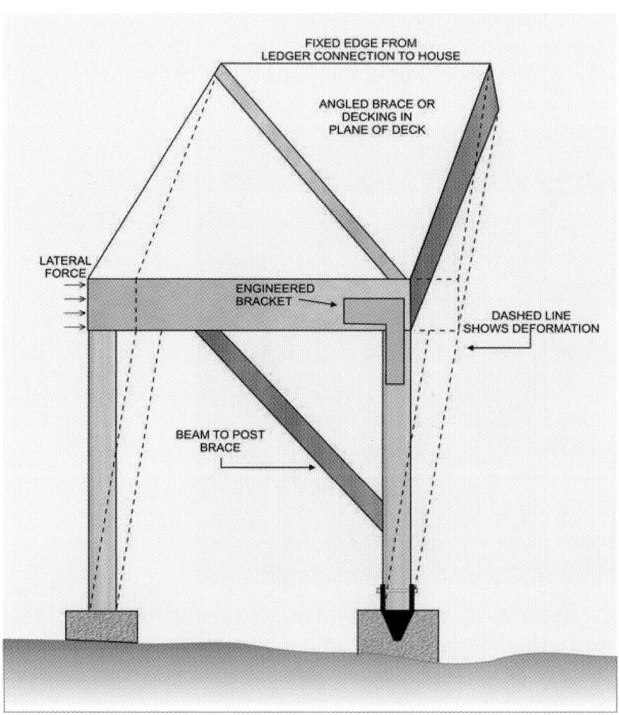

Example 4-9: Three possible methods of lateral bracing are shown in this example; however, none of them are referenced by the IRC and all must be approved as "alternatives." When one or more sides of the deck are in a fixed position, such as a ledger connection, lateral resistance can be achieved by eliminating the possibility for sway or movement in the deck surface. A square can be deflected into a parallelogram shape, as shown by the dashed lines in the example, but a triangle is inherently more rigid. This is why trusses are made of triangles. The addition of angled braces or angled decking in the plane of the deck or the post/beam area will create a triangle shape and resist movement of the deck.

Another means of resisting lateral loads is through a rigid connection created by an engineered bracket. A rigid connection locks the 90-degree (1.57 rad) angle in one corner of the square, thus resisting movement of the deck. A typical manufactured post cap does not provide a rigid connection, so this type of lateral resistance must be specifically engineered. Generally speaking, it is infeasible or impractical to create a sufficiently rigid connection in a wood-framed structure, due to the lack of stiffness provided by typical connections used in wood structures. Under design loading, the bracket may indeed maintain its 90-degree (1.57 rad) shape, but the wood at the connection location may deform or fail.

Example 4-10: Two metal straps were used on this deck in the plane of the decking as a means to create triangles that will lock the deck surface from swaying.

Example 4-11: Similar to Example 4-10, these 2 by 6 cross braces lock the plane of the posts and beams, and will effectively prevent any lateral movement.

R301.1.3 Engineered design. When a building of otherwise conventional construction contains structural elements exceeding the limits of Section R301 or otherwise not conforming to this code, these elements shall be designed in accordance with accepted engineering practice. The extent of such design need only demonstrate compliance of nonconventional elements with other applicable provisions and shall be compatible with the performance of the conventional framed system. Engineered design in accordance with the *International Building Code* is permitted for all buildings and structures, and parts thereof, included in the scope of this code.

Discussion: Many deck designs will require the use of "alternative" designs in order to satisfy portions of the complete load path, ones that must be approved specifically by the local building official. If you are not able to find pre-engineered documents, like those provided in the appendix of this book, or test results to obtain approval from the local building official, your other option is site-specific engineered design. Before employing a design professional, it is always prudent to consult with the local building department. Many jurisdictions adopt their own policies and requirements for what structures must be engineered. Likewise, the registration and licensure required for design professionals to perform engineering is regulated by the state authority.

This section allows the use of both prescriptive designs from the IRC concurrently with engineered design. For example, a deck may be designed utilizing IRC design provisions for the ledger connection, alternative engineered documents for beam spans and site-specific engineering for the lateral resistance. In whatever combination, if the load path is complete and can be verified, the IRC structural design requirement is satisfied.

R502.2 Design and construction. Floors shall be designed and constructed in accordance with the provisions of this chapter, Figure R502.2 and Sections R317 and R318 or in accordance with AF&PA/NDS.

Discussion: Chapter 5 of the IRC regulates the construction of floors. Most of the structural design provisions that can still be used for post-frame construction of decks are contained in this IRC chapter, as a deck is a floor. Although, a deck is an unconventional floor—one generally exposed to the weather and without braced walls or horizontal diaphragms for lateral resistance. This section basically states that floors must be designed in accordance with Chapter 5 of the IRC (see

Part 5 of this chapter) and the material-decay-resistant requirements in the IRC (previously discussed in Part 1 of this chapter).

The last statement of this section allows the use of the AF&PA/NDS (*National Design Specification for Wood Construction*). This document is the referenced standard for engineering wood-frame structures, and its use is more appropriate for a design professional.

R301.2 Climatic and geographic design criteria. Buildings shall be constructed in accordance with the provisions of this code as limited by the provisions of this section. Additional criteria shall be established by the local *jurisdiction* and set forth in Table R301.2(1).

TABLE R301.2(1)
CLIMATIC AND GEOGRAPHIC DESIGN CRITERIA

GROUND SNOW LOAD	WIND DESIGN		SEISMIC DESIGN CATEGORY[f]	SUBJECT TO DAMAGE FROM			WINTER DESIGN TEMP[e]	ICE BARRIER UNDERLAYMENT REQUIRED[h]	FLOOD HAZARD[g]	AIR FREEZING INDEX[i]	MEAN ANNUAL TEMP[j]
	Speed[d] (MPH)	Topographic effects[k]		Weathering[a]	Frost line depth[b]	Termite[c]					

For SI: 1 pound per square foot = 0.0479 kPa, 1 mile per hour = 0.447 m/s.

a. Weathering may require a higher strength concrete or grade of masonry than necessary to satisfy the structural requirements of this code. The weathering column shall be filled in with the weathering index (i.e., "negligible," "moderate" or "severe") for concrete as determined from the Weathering Probability Map [Figure R301.2(3)]. The grade of masonry units shall be determined from ASTM C 34, C 55, C 62, C 73, C 90, C 129, C 145, C 216 or C 652.

b. The frost line depth may require deeper footings than indicated in Figure R403.1(1). The jurisdiction shall fill in the frost line depth column with the minimum depth of footing below finish grade.

c. The jurisdiction shall fill in this part of the table to indicate the need for protection depending on whether there has been a history of local subterranean termite damage.

d. The jurisdiction shall fill in this part of the table with the wind speed from the basic wind speed map [Figure R301.2(4)].Wind exposure category shall be determined on a site-specific basis in accordance with Section R301.2.1.4.

e. The outdoor design dry-bulb temperature shall be selected from the columns of $97^1/_2$-percent values for winter from Appendix D of the International Plumbing Code. Deviations from the Appendix D temperatures shall be permitted to reflect local climates or local weather experience as determined by the building official.

f. The jurisdiction shall fill in this part of the table with the seismic design category determined from Section R301.2.2.1.

g. The jurisdiction shall fill in this part of the table with (a) the date of the jurisdiction's entry into the National Flood Insurance Program (date of adoption of the first code or ordinance for management of flood hazard areas), (b) the date(s) of the Flood Insurance Study and (c) the panel numbers and dates of all currently effective FIRMs and FBFMs or other flood hazard map adopted by the authority having jurisdiction, as amended.

h. In accordance with Sections R905.2.7.1, R905.4.3.1, R905.5.3.1, R905.6.3.1, R905.7.3.1 and R905.8.3.1, where there has been a history of local damage from the effects of ice damming, the jurisdiction shall fill in this part of the table with "YES." Otherwise, the jurisdiction shall fill in this part of the table with "NO."

i. The jurisdiction shall fill in this part of the table with the 100-year return period air freezing index (BF-days) from Figure R403.3(2) or from the 100-year (99%) value on the National Climatic Data Center data table "Air Freezing Index- USA Method (Base 32°)" at www.ncdc.noaa.gov/fpsf.html.

j. The jurisdiction shall fill in this part of the table with the mean annual temperature from the National Climatic Data Center data table "Air Freezing Index-USA Method (Base 32°F)" at www.ncdc.noaa.gov/fpsf.html.

k. In accordance with Section R301.2.1.5, where there is local historical data documenting structural damage to buildings due to topographic wind speed-up effects, the jurisdiction shall fill in this part of the table with "YES." Otherwise, the jurisdiction shall indicate "NO" in this part of the table.

Discussion: The IRC, like other I-Codes, is written with the intent of providing uniform and consistent construction practices within the United States and other places where the code is adopted. However, the United States varies too much in climatic conditions for complete uniformity to occur from coast to coast. In the attempt to provide standards for construction that will reflect the intent and purpose of the IRC, consideration must be made as to where the structure will be located. Mother Nature is not the same beast when at the coast of Florida, mountains of Colorado, humid forests of Oregon or the wind-swept regions of Kansas. Proper design and construction must consider the climate and geography in which the deck is to be constructed.

Table R301.2(2) provides a format and pretitled list of criteria that must be locally determined. From these criteria, many requirements in the IRC will vary. Luckily for deck construction, only a handful of these criteria are applicable. Familiar to most is the locally determined frost depth, which is the minimum depth for which foundation systems must extend below the surface of the earth. Also to be considered in the design of the structure and material selection is the ground snow load, the wind speed and the need for termite protection. These criteria can only be provided by the local building official, and should be determined before construction begins.

R301.5 Live load. The minimum uniformly distributed live load shall be as provided in Table R301.5.

TABLE R301.5
MINIMUM UNIFORMLY DISTRIBUTED LIVE LOADS
(in pounds per square foot)

USE	LIVE LOAD
Attics without storage[b]	10
Attics with limited storage[b, g]	20
Habitable attics and attics served with fixed stairs	30
Balconies (exterior) and decks[e]	40
Fire escapes	40
Guardrails and handrails[d]	200[h]
Guardrail in-fill components[f]	50h
Passenger vehicle garages[a]	50[a]
Rooms other than sleeping room	40
Sleeping rooms	30
Stairs	40[c]

For SI: 1 pound per square foot = 0.0479 kPa, 1 square inch = 645 mm^2,
1 pound = 4.45 N.

a. Elevated garage floors shall be capable of supporting a 2,000-pound load applied over a 20-square-inch area.

b. Attics without storage are those where the maximum clear height between joist and rafter is less than 42 inches, or where there are not two or more adjacent trusses with the same web configuration capable of containing a rectangle 42 inches high by 2 feet wide, or greater, located within the plane of the truss. For attics without storage, this live load need not be assumed to act concurrently with any other live load requirements.

c. Individual stair treads shall be designed for the uniformly distributed live load or a 300-pound concentrated load acting over an area of 4 square inches, whichever produces the greater stresses.

d. A single concentrated load applied in any direction at any point along the top.

e. See Section R502.2.2 for decks attached to exterior walls.

f. Guard in-fill components (all those except the handrail), balusters and panel fillers shall be designed to withstand a horizontally applied normal load of 50 pounds on an area equal to 1 square foot. This load need not be assumed to act concurrently with any other live load requirement.

g. For attics with limited storage and constructed with trusses, this live load need be applied only to those portions of the bottom chord where there are two or more adjacent trusses with the same web configuration capable of containing a rectangle 42 inches high or greater by 2 feet wide or greater, located within the plane of the truss. The rectangle shall fit between the top of the bottom chord and the bottom of any other truss member, provided that each of the following criteria is met.

1. The attic area is accessible by a pull-down stairway or framed in accordance with Section R807.1.
2. The truss has a bottom chord pitch less than 2:12.
3. Required insulation depth is less than the bottom chord member depth.

The bottom chords of trusses meeting the above criteria for limited storage shall be designed for the greater of the actual imposed dead load or 10 psf, uniformly distributed over the entire span.

h. Glazing used in handrail assemblies and guards shall be designed with a safety factor of 4. The safety factor shall be applied to each of the concentrated loads applied to the top of the rail, and to the load on the in-fill components. These loads shall be determined independent of one another, and loads are assumed not to occur with any other live load.

Discussion: So far all of the IRC sections provided in this part of Chapter 4 have only discussed design methods and the requirement for a complete load path, yet there has been no mention of what the loads are. The dead loads, as defined previously, are relatively easy to determine: they are simply the weight of the structure itself based on the materials employed. Dead loads also include permanent equipment and features, such as hot tubs, built-in barbeques and other such amenities. Most span tables have columns for dead loads of 10 pounds per square foot and 20 pounds per square foot [see the discussion of Table R502.3.1(2)].

Live loads, however, vary depending on the structural component in question. In general, live loads result from the use or occupancy of the structure. Table R301.5, provides live load design criteria for many of the common structural components found in typical residential construction. This discussion will review two of the components in the table, decks and exterior balconies. Guard and handrail live loads are discussed in Chapter 7 of this book and stair and stair tread live loads in Chapter 6.

The live load design requirement for all decks, including balconies and regardless of which room they are accessed from, is 40 pounds per square foot (1.9 kPa) of deck area. When using the joist span tables, either in the IRC or from other sources, it is important to verify that the tables are based on a 40 pounds per square foot (1.9 kPa) uniformly distributed live load or greater. A uniformly distributed load assumes just that, that the loads are distributed evenly throughout the area, and applied concurrently with one another.

Part Four: Ledgers

R703.7.3 Lintels. Masonry veneer shall not support any vertical load other than the dead load of the veneer above. Veneer above openings shall be supported on lintels of noncombustible materials. The lintels shall have a length of bearing not less than 4 inches (102 mm). Steel lintels shall be shop coated with a rust-inhibitive paint, except for lintels made of corrosion-resistant steel or steel treated with coatings to provide corrosion resistance. Construction of openings shall comply with either Section R703.7.3.1 or 703.7.3.2.

Discussion: When a deck ledger is connected through a brick veneer and to the band joist within the structure, a load will in almost all cases still be applied to the brick veneer. The only vertical load that can be imposed on anchored masonry veneer is the weight of the bricks themselves (see Part 7 of Chapter 2 for a greater understanding of masonry veneer). Much like a single 2 by 4 temporary brace may bend out horizontally under excess vertical loading, so can a brick veneer wall that has not been designed or intended to support a vertical load. The wire-tie requirement in the IRC for anchoring veneer to an exterior wall is neither sufficient nor intended to resist the horizontal buckling of the veneer under additional load conditions. Of course much like a 2 by 4 temporary brace, the buckling is directly related to the magnitude of the vertical load and the total height of the brick veneer wall. A short veneer wall of only a couple feet with a minimal vertically imposed load could very well function properly as a structural member, but not without an engineer's evaluation and the local building official's approval of the alternative.

In other configurations, however, the implications of this section may make it financially, structurally and aesthetically difficult to attach a ledger over a brick veneer (see Example 4-12).

A ledger can, however, be attached through the brick to resist laterally imposed loads on the deck. One prescriptive method of achieving this lateral connection at the ledger connection is provided later in this chapter in the discussion of IRC Section R502.2.2.3. In new construction applications, the ledger or joists can be connected directly to the wood framing and the brick can be installed around the joists (see Example 4-2 in Part 1 of this chapter). This method, however, creates difficulty for future joist replacement and requires the use of decay-resistant wood material as discussed in Part 1 of this chapter.

Example 4-12: An engineered design was employed to correct this deck whose ledger was imposing vertical loads on the brick veneer. While this was an unplanned repair to the construction, it's one of the options that could have been used in the original design. The remediation consisted of a beam, post and foundation bracket system to carry the vertical loads from the ledger.

R502.2.2 Decks. Where supported by attachment to an exterior wall, decks shall be positively anchored to the primary structure and designed for both vertical and lateral loads as applicable. Such attachment shall not be accomplished by the use of toenails or nails subject to withdrawal. Where positive connection to the primary building structure cannot be verified during inspection, decks shall be self-supporting. For decks with cantilevered framing members, connections to exterior walls or other framing members, shall be designed and constructed to resist uplift resulting from the full live load specified in Table R301.5 acting on the cantilevered portion of the deck.

Discussion: The typical connection of a deck to an existing structure is through the use of a ledger board attached to a wall, floor band (rim) joist or directly to the foundation. No matter the means of deck connection to a primary structure, it must be capable of resisting both vertical and lateral (horizontal) loads. This connection cannot be achieved through the use of toenails or nails installed in a manner where they will be subject to withdrawal (forces in the direction of the shank of the nail). Nails provide significantly greater resistance in their shear strength (bearing on the wood perpendicular to the shank of the nail); however, in a typical deck connection to a primary structure this orientation of nails is not usually achieved.

The connection of a deck to a primary structure must be visible for inspection. In a connection to a floor system that is concealed on the inside, it cannot be verified that a lag screw has penetrated a band joist sufficiently or that a proper band joist even exists (see Examples 4-13 and 4-14). If the load path cannot be substantially verified within the structure, the deck must be supported independently.

When a deck is constructed with cantilevered floor members, the opposite end of cantilevered joists must be capable of resisting the uplift forces in addition to the vertical and lateral forces. Deck cantilevers and these uplift forces are discussed in further detail later in this chapter.

Example 4-13: Exterior walls and band joists (rim joists) of existing structures may contain a variety of construction components, such as pipes, ducts and electrical cables. In this photo the rim/band joist material has been removed to fit the turn of the ductwork and the plumbing drain is against the backside of the band joist. Tightening a lag screw into a duct, pipe or electrical cable will not only compromise the structural integrity of the connection, but may also create damage to existing systems within the existing structure.

Example 4-14: Unrecognizable from the exterior of the dwelling, an open web floor system may not contain a band joist at all. Unless verified from inside, a lag screw could presumably tighten against only the ribbon strip of $^1/_2$-inch (12.7 mm) thick wall sheathing outside the ends of these joists, which is not a sufficient connection at all.

R502.2.2.1 Deck ledger connection to band joist. For decks supporting a total design load of 50 pounds per square foot (2394 Pa) [40 pounds per square foot (1915 Pa) live load plus 10 pounds per square foot (479 Pa) dead load], the connection between a deck ledger of pressure-preservative-treated Southern Pine, incised pressure-preservative-treated Hem-Fir or *approved* decay-resistant species, and a 2-inch (51 mm) nominal lumber band joist bearing on a sill plate or wall plate shall be constructed with $^1/_2$-inch (12.7 mm) lag screws or bolts with washers in accordance with Table R502.2.2.1. Lag screws, bolts and washers shall be hot-dipped galvanized or stainless steel.

TABLE R502.2.2.1
FASTENER SPACING FOR A SOUTHERN PINE OR HEM-FIR DECK LEDGER AND A 2-INCH NOMINAL SOLID-SAWN SPRUCE-PINE-FIR BAND JOIST[c, f, g]
(Deck live load = 40 psf, deck dead load = 10 psf)

JOIST SPAN	6′and less	6′1″ to 8′	8′1″ to 10′	10′1″ to 12′	12′1″ to 14′	14′1″ to 16′	16′1″ to 18′
Connection details	On-center spacing of fasteners[d,e]						
$^1/_2$ inch diameter lag screw with $^{15}/_{32}$ inch maximum sheathing[a]	30	23	18	15	13	11	10
$^1/_2$ inch diameter bolt with $^{15}/_{32}$ inch maximum sheathing	36	36	34	29	24	21	19
$^1/_2$ inch diameter bolt with $^{15}/_{32}$ inch maximum sheathing and $^1/_2$ inch stacked washers[b, h]	36	36	29	24	21	18	16

a. The tip of the lag screw shall fully extend beyond the inside face of the band joist.
b. The maximum gap between the face of the ledger board and face of the wall sheathing shall be $^1/_2$".
c. Ledgers shall be flashed to prevent water from contacting the house band joist.
d. Lag screws and bolts shall be staggered in accordance with Section R502.2.2.1.1.
e. Deck ledger shall be minimum 2x8 pressure-preservative-treated No.2 grade lumber, or other approved materials as established by standard engineering practice.
f. When solid-sawn pressure-preservative-treated deck ledgers are attached to engineered wood products (structural composite lumber rimboard or laminated veneer lumber), the ledger attachment shall be designed in accordance with accepted engineering practice.
g. A minimum 1 by 9-$^1/_2$ Douglas Fir laminated veneer lumber rimboard shall be permitted in lieu of the 2-inch nominal band joist.
h. Wood structural panel sheathing, gypsum board sheathing or foam sheathing not exceeding 1 inch in thickness shall be permitted. The maximum distance between the face of the ledger board and the face of the band joist shall be 1 inch.

Discussion: This helpful IRC section and table is new to the 2009 IRC as well as the following subsections. The most common connection of a deck to an existing structure is a ledger board bolted to an existing band joist. To provide requirements for the most prevalent ledger board materials, this section is applicable to pressure-preservative-treated Southern Pine, pressure-preservative-treated and incised Hem-Fir or other *approved* decay-resistant species of lumber, as discussed in Part 1 of this chapter. For prescriptive use of Table R502.2.2.1, the band joist inside the existing structure must be verified as 2-inch (51 mm) nominal solid sawn lumber of a select variety of species. This band joist must bear directly on a foundation or wall sill; it cannot be a band joist connected to cantilevered floor joists. More specific than the general requirement for fasteners in preservative-treated materials, discussed earlier in this chapter, the ½-inch (12.7 mm) lag screws or bolts for this application do not apply to the exception to Section R317.3.1. Bolts used for compliance with this table must be either hot-dipped galvanized or stainless steel. The intent of this section is that the hot-dipped galvanization be in accordance with ASTM A 153, as discussed in Part 2 of this chapter.

When using this table, the on-center spacing of the joists is not a concern, as that is related to the span of the decking material, not the overall load applied to the ledger. What is important is the overall span of the joists, as this is related to the overall area supported and the total live load that must be resisted at the ledger. In many deck designs, different portions of the ledger will require different on-center spacing of the bolts or lag screws as a result of differing spans of the joists it supports. Beyond the basic understanding of the rows and columns of the table, there are also many construction details provided in the notes that must be satisfied by the installation.

Note a reinforces the requirement for a visible connection that can be verified, as required by the parent section of this table. Rather than simply specifying a minimum length lag screw that could not be verified after installation, the note directly requires the lag to penetrate the complete thickness of the band joist material.

Notes b and h allow the ledger board to be held off the face of the band joist up to a maximum of 1 inch (25 mm). In a bolted connection, the ledger may be spaced from the band joist by any combination of ½-inch (12.7 mm) thickness of stacked washers or structural, foam or gypsum sheathing, up to the 1-inch (25 mm) maximum. It is important to realize that these notes only apply to bolted ledger connections and not lag screw connections. A lag screw connection would require the ledger to be placed directly against structural wall sheathing or the band joist itself.

Reinforcing Section R703.8, discussed later in this part, note c reiterates the necessity to flash the ledger/band joist connection from water intrusion. Note d refers to Section R502.2.2.1.1 for the placement of lag screws and bolts and is reviewed in the following discussion.

Notes e, f and g are all concerned with the ledger and band joist material. The ledger material must be a minimum of 2 by 8 nominal No. 2 grade lumber that is pressure-preservative treated or other approved decay-resistant material, such as naturally durable wood. An exception is provided to the nominal band joist requiremnet in Section R502.2.2.1 and allows, specifically, a Douglas-Fir laminated veneer lumber rimboard of at least 1-inch by $9^1/_2$-inches (25 mm by 241 mm) dimensional. Other engineered lumber products could be used if approved by the building official and installed in accordance with the manufacturer's installation instructions.

While this new table does provide prescriptive ledger connection criteria, it is also very specific. Using this table for a ledger installation must be completed with close attention to the limitations and specific details within the table and the related sections.

R502.2.2.1.1 Placement of lag screws or bolts in deck ledgers. The lag screws or bolts shall be placed 2 inches (51 mm) in from the bottom or top of the deck ledgers and between 2 and 5 inches (51 and 127 mm) in from the ends. The lag screws or bolts shall be staggered from the top to the bottom along the horizontal run of the deck ledger.

Discussion: The locations along the ledger where lag screws or bolts can be installed are intended to maintain the structural integrity of the ledger material from the forces between the bolt and the wood member. Reflecting the allowable drilling provisions of joists, discussed later in this chapter, the lag screws and bolts must be a minimum of 2 inches (51 mm) in from the top and bottom edge of the ledger material. To resist cracking along the grain of the lumber at the ends of the ledger and to limit the extension of a ledger beyond the last connection, the lag screws or bolts at the end of the ledger must be between 2 and 5 inches (51 and 127 mm) from each end of the ledger board. To minimize cupping and rotation of the ledger board, the connections must be staggered in an alternating pattern from the top to the bottom edges along the length of the ledger (see Example 4-15).

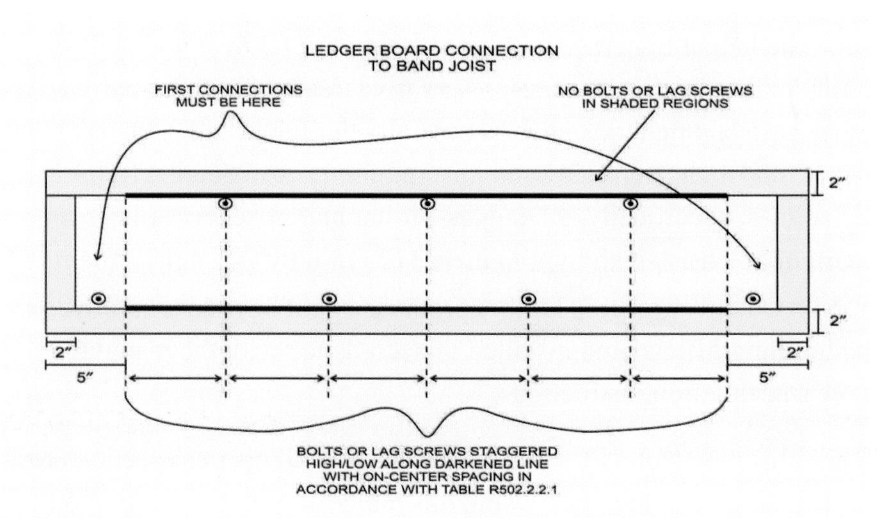

Example 4-15

R502.2.2.2 Alternate deck ledger connections. Deck ledger connections not conforming to Table R502.2.2.1 shall be designed in accordance with accepted engineering practice. Girders supporting deck joists shall not be supported on deck ledgers or band joists. Deck ledgers shall not be supported on stone or masonry veneer.

Discussion: This short section of only three statements produces powerful limitations to deck design, and prohibits many previously somewhat common installation practices. The first statement is simple: Ledger connections not conforming to Table R502.2.2.1 must be an engineered design. However, this statement is nothing new. As required by Section R301.1.3, and previously discussed in this chapter, designs not conforming to the method prescribed by the IRC must be based on engineering principles and are subject to approval as an alternative. Ledger connections utilizing alternative connection details, and ledgers connected to concrete foundations or wall studs, are examples of connections that would require an alternative means, based on engineering principles, as evidence that all loads are sufficiently transferred to the supporting soil.

Similar to this first statement, the last statement is also reiterating a provision provided in greater detail elsewhere in the IRC. Section R703.7.3, discussed previously, clearly states that stone or masonry veneers cannot support any additional load other than the dead load of the veneer above.

The significant statement in this section is the second. A structural member supporting other deck joists, such as a double joist supporting a stair landing, cannot bear at a ledger or band joist. A joist or beam supporting other joists creates a larger load to support at both ends. This concentrated load cannot be supported by a deck ledger or band joist, as it exceeds the engineering calculations that produced the fastening requirements in Table R502.2.2.1. A beam, girder or joist supporting other members must be extended into the framed structure and bear directly on top of its supporting member (see Example 4-16). If not extended into the existing framed structure, the concentrated loads must otherwise be transferred to undisturbed supporting soil, such as with the use of a separate post and/or pier, or an engineered ledger connection.

This does not, however, prohibit a beam from being supported in a different location than the ledger or band joist through the use of a joist hanger, provided the loads have been calculated and supported in accordance with engineering principles.

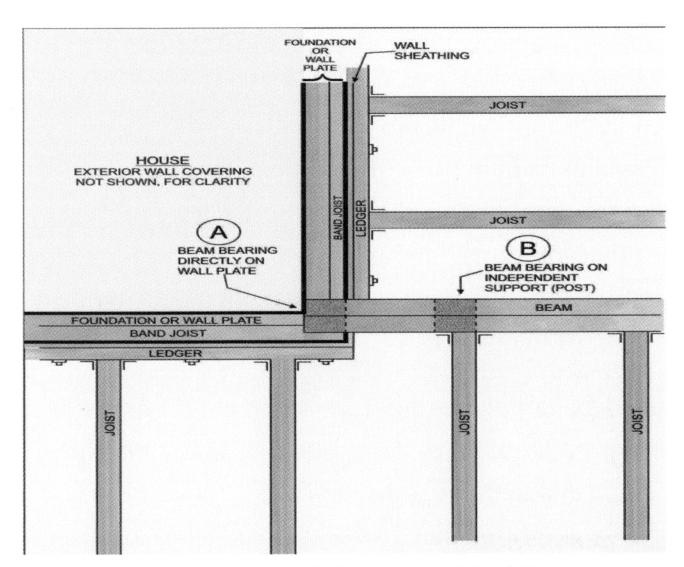

Example 4-16: This example is a typical deck frame constructed around the outside corner of an existing structure. The beam used to carry the joists on one side of the deck cannot be attached to the ledger or band joist. Option "A" shows the beam extending into the existing structure and resting on the foundation or wall plate below, while "B" depicts an independent support provided underneath the beam, such as a post and/or pier.

R502.2.2.3 Deck lateral load connection. The lateral load connection required by Section R502.2.2 shall be permitted to be in accordance with Figure R502.2.2.3. Hold-down tension devices shall be installed in not less than two locations per deck, and each device shall have an allowable stress design capacity of not less than 1500 pounds (6672 N).

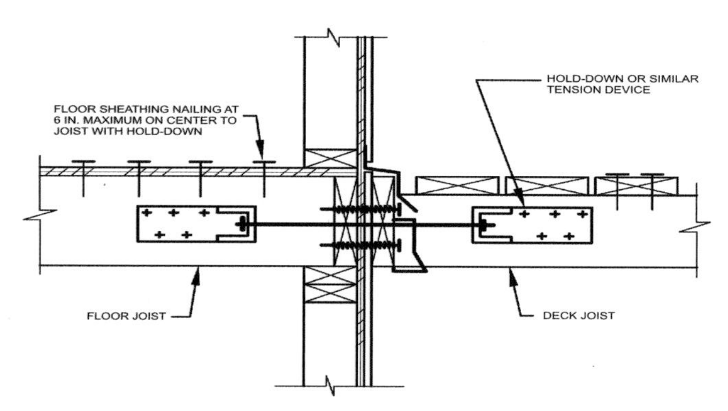

IRC Figure 502.2.2.3: Deck Attachment for Lateral Loads

Discussion: Section R502.2.2, previously discussed, requires a ledger connection to resist both vertical and lateral (horizontal) loads. In some instances, the band joist in an existing structure may not be securely connected to the rest of the structure in a manner adequate to resist lateral loads from specific deck designs. In these instances, or when additional lateral resistance is necessary beyond that provided by the ledger connection in Table 502.2.2.1, this connection detail can be used to tie the deck joists to the existing floor joists, essentially bypassing the ledger and band joist all together. This detail is not required by the IRC, but rather "permitted," and it is essentially a preapproved means to resist lateral loads that meets the most severe lateral load conditions anticipated for decks within the scope of the IRC. The overall design of the deck and the condition of the existing structure will determine the necessity for this additional lateral load resistance or not. This is a professional judgment that will need to be made on a deck by deck basis during the approval of the project.

All the specifics of this IRC detail are only feasible in new construction applications, as otherwise it would require removal of many interior finishes. However, since this detail is not specifically required, it can still provide additional lateral resistance to decks attached to existing structures where the floor sheathing is covered and cannot be nailed or verified in accordance with this section.

In two or more locations of the deck structure, ideally near the outside edges of the deck, a hold-down device can be installed on the deck joist and on the existing structures floor joist. These hold-downs are then connected together, through the band joist and ledger, with a bolt. The floor sheathing can be nailed at 6 inches (152 mm) on-center at the existing structures joists where the hold-downs are located. Each hold-down should be capable of resisting a 1,500 pound (6672 N) load. This detail does not offer a method of installation when the floor joists are running parallel to the wall. In this case it would be reasonable to install blocking between the joists at the same spacing as the floor joists and nail the floor sheathing to the blocking 6 inches (152 mm) on center with the hold-down connected to the blocking.

Deck Connections at Existing Floor Cantilever Locations

Discussion: Quite often in residential deck construction, a deck design may encompass a cantilevered floor section of the dwelling. Section R502.2.2.1 for ledger connections to band joists requires the band joist to be directly bearing on a sill or wall plate below. At floor cantilevers, the band joist is only connected through end-nailing into the ends of the cantilevered floor joists, rather than being directly supported. For this condition a deck ledger cannot be fastened to a band joist using Table R502.2.2.1. In the case that an alternative design creates a sufficient connection between the end of the floor joists and the band joist, the concentrated loads placed on the ends of the cantilevered joists may create excessive bending, deformation and deflection, such that the structural integrity of the cantilevered joists are compromised. Generally speaking, careful consideration by a design professional is required to sufficiently make this connection.

In lieu of an engineered design, there are other methods that may be used to design a deck around a cantilevered floor section.

One method of framing a deck encompassing a cantilever is to construct the deck into the cantilevered floor. By removing the existing wall covering, soffit material, sheathing, band joist and blocking above the bearing wall, the deck joists can extend beside the existing cantilevered floor joists and bear directly on the wall or foundation plate (see Example 4-17). The direct bearing on the plate below bypasses the cantilever entirely and provides adequate support of the vertical loads. To resist lateral loads (horizontal) the deck joists can be placed against the existing joists and nailed together. This orientation of the nails will resist lateral loads through the shear strength of the nails (bearing on wood perpendicular to the shank) and would not be considered "subject to withdrawal" as prohibited by Section R502.2.2.

When exposing an existing wall or floor cavity that is part of the buildings insulated thermal envelope, the cavity should be filled with insulation prior to closure.

Example 4-17: These deck joists are extended into the existing structure for direct bearing on the wall plate below. This method is a very effective way to connect a deck structure to an existing structure at a cantilever location.

> **Side Note:** The IRC allows joists to be notched vertically at their bearing points. This allowance can be very helpful in adjusting the height of the top of the deck joists when extended into a cantilever. For example, a 2 by 12 deck joist could be notched such that it fits into an existing 2 by 10 floor cavity, at the bearing location. See the discussion of Section R502.8 later in this chapter.

A second method of constructing a deck frame at an existing cantilevered floor is to frame around the cantilever. Using a combination of beams, the load path for the deck joists can transfer to a beam at the face of the cantilever, in lieu of a ledger. This beam can then be supported by two additional beams at each end, extending perpendicular to the face of the cantilever (see Example 4-18). As discussed under Section R502.2.2.2, these beams could not be supported by the ledger or band joist. Provisions in the IRC for framing at floor openings can be approved for this application in accordance with the limitations provided in Section R502.10 (see Part 5 of this chapter).

A slight variation of this method may also be used; however, it may not be desirable to the deck owners. Rather than a beam in place of a ledger and supported by additional beams, a dropped beam can be placed below the joists that are located at the cantilever (see Example 4-19). The disadvantage of this method is the requirement for two or more additional posts and piers to support the beam. The beam could be held off the ledger a maximum distance as determined by Section R502.3.3, discussed in Part 5 of this chapter. This "reverse" cantilever coupled with the distance of the existing cantilever may place the additional piers outside of the backfill zone of the existing foundation where undisturbed soil can provide support (see Chapter 3).

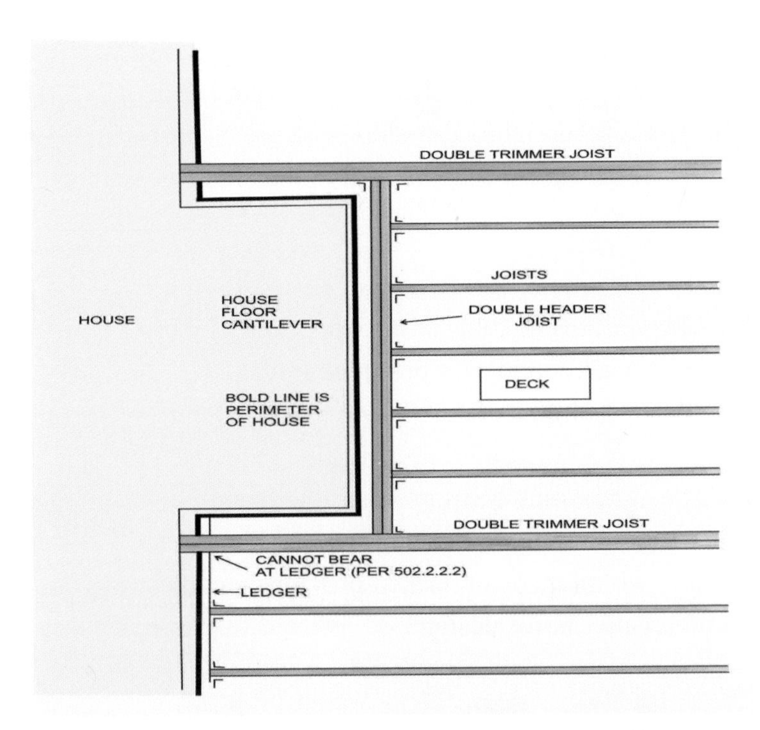

Example 4-18: This illustrations shows how a series of beams can be used to bypass a ledger connection to an existing floor cantilever. All members would need to be properly sized based on the actual loads received.

Example 4-19: A dropped beam was installed under this deck rather than trying to support the deck by the cantilevered floor. The beam was moved out from the face of the cantilevered floor, allowing its posts to bear more than 4 feet (1219 mm) from the existing foundation, in undisturbed soil. This "reverse cantilever" will keep the footings from this beam from bearing in the backfill region of the existing foundation.

Another method of deck connection at existing floor cantilevers is to drop the deck frame beneath the cantilever and connect the ledger to the wall studs or foundation beneath the existing floor (see Example 4-20). Difficulties in this method, however, may not be desirable for the final design of the deck. If a door from the existing structure accesses the deck, a stairway would likely need to be provided of up to two risers. In the case that this door is the designated required "egress door" in accordance with IRC Section R311.2, and more than one riser was needed, a fully compliant landing would need to be constructed on top of the deck surface. Details of landings at doors are discussed in Chapter 2 of this book.

This method would also require an alternative design for the connection of the ledger to wall studs or a foundation, as there are no prescriptive means for that connection provided in the IRC. A design professional would be required to detail this connection on the plan that is submitted for approval to the building official. Once a need arises for an engineered design, it may be more desirable to simply have the connection of the ledger directly to the cantilevered band joist evaluated. The first two methods described previously can usually be completed without an alternative design.

Example 4-20: The connection of this ledger to the foundation eliminates the concerns of a connection to the cantilever, yet requires approval from the building official as an "alternative" design for bolting into the foundation.

R703.8 Flashing. *Approved* corrosion-resistant flashing shall be applied shingle-fashion in a manner to prevent entry of water into the wall cavity or penetration of water to the building structural framing components. Self-adhered membranes used as flashing shall comply with AAMA 711. The flashing shall extend to the surface of the exterior wall finish. *Approved* corrosion-resistant flashings shall be installed at all of the following locations:

1. Exterior window and door openings. Flashing at exterior window and door openings shall extend to the surface of the exterior wall finish or to the water-resistive barrier for subsequent drainage.
2. At the intersection of chimneys or other masonry construction with frame or stucco walls, with projecting lips on both sides under stucco copings.
3. Under and at the ends of masonry, wood or metal copings and sills.
4. Continuously above all projecting wood trim.
5. Where exterior porches, decks or stairs attach to a wall or floor assembly of wood-frame construction.
6. At wall and roof intersections.
7. At built-in gutters.

Discussion: Inhibiting water damage to a structure from intrusion into the building envelope is of primary importance. However, this can be difficult to achieve with the presence of windows, doors, architectural details, joints of differing materials and, of course, deck ledger attachments. This section describes the general function and purpose of flashing and provides a list of seven specific locations where its use is mandatory.

The fifth item in the list specifically requires flashing where decks, porches or stairs attach to a wood-frame structure, but there are no specific details provided for how this is to be accomplished (see Example 4-21). There are, however, some basic criteria that must be met. Flashing must be installed in "shingle fashion," which basically means that as gravity pulls water toward the earth, it will always shed onto the outer surface of the material below—just like shingles lap over the shingles below. Regardless of the method chosen to flash a ledger, it must be shingle fashion and it must keep the water out. If water gets in, not only is it is a code violation, but it may lead to a situation where serious damage may occur to the structure (see Example 4-22).

In regard to what material can be used as flashing, there is no definitive requirement for metal or anything else. Flashing can be of any material, provided it complies with the two requirements of this section: approved and corrosion resistant. In considering these adjectives, "approved" refers to approval by the local building official. Corrosion resistance is an issue that needs to be discussed. The corrosion resistance of flashing material is not based solely on the flashing material itself, but also on the location in which it is placed. Materials usually considered corrosion resistant, such as galvanized steel or aluminum, lose some of that capability when placed in contact with some dissimilar metals, such as the copper in pressure-preservative-treated wood (see the discussion of Section R317.3 in Part 1 of this chapter for details). For preservative-treated wood ledgers, only products like copper, vinyl or self-adhering polymer-modified bitumen, or other products that do not react with copper, can be considered corrosion-resistant.

Example 4-21: This stair stringer was cut into the siding and attached to the structure with no flashing. While this type of installation is not recommended, it could have been compliant if a flashing was installed to shed the water to the outer surface of the stringer. The simpler choice would be to gap the stair stringer from the wall to allow drainage and drying of the materials.

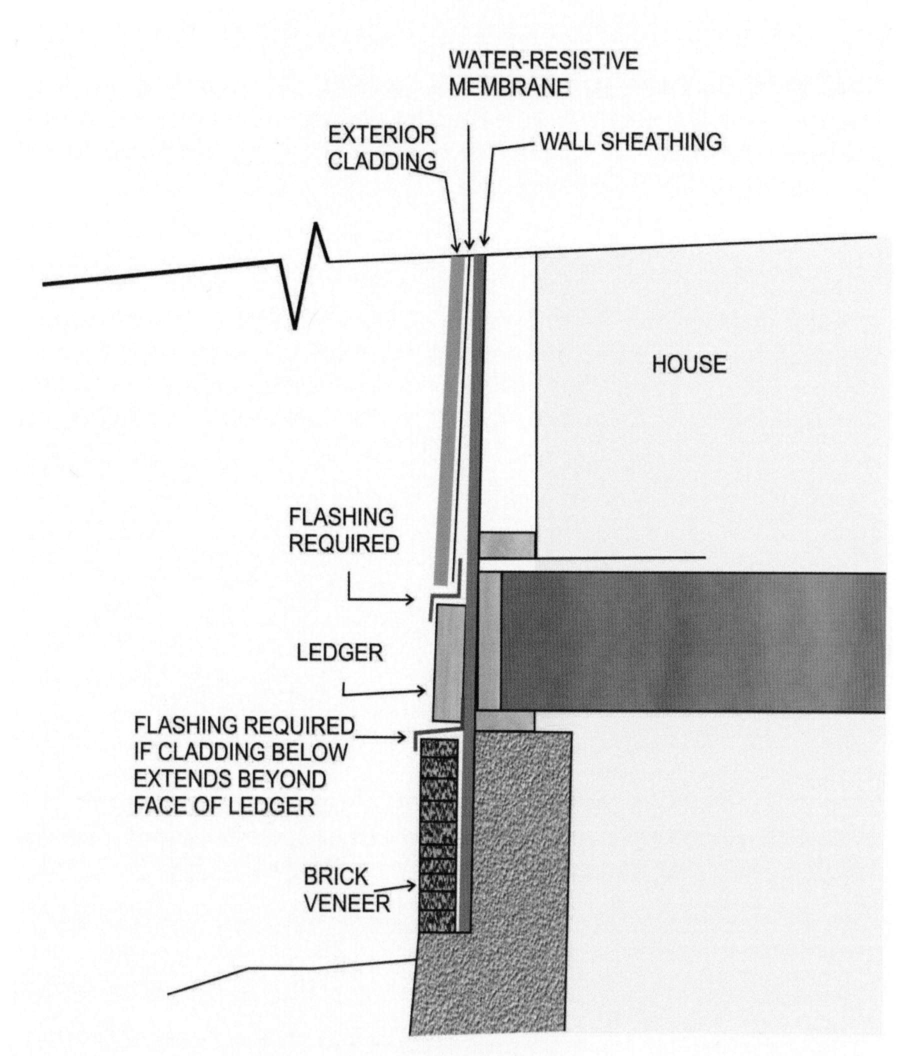

Example 4-22: As mentioned in the discussion, the purpose of the flashing is to keep water on the face of the ledger and exterior cladding, and not allow it to enter the building cavity. This illustration details one such way that can be achieved.

Part Five: Framing Criteria

R502.3 Allowable joist spans. Spans for floor joists shall be in accordance with Tables R502.3.1(1) and R502.3.1(2). For other grades and species and for other loading conditions, refer to the AF&PA Span Tables for Joists and Rafters.

R502.3.2 Other floor joists. Table R502.3.1(2) shall be used to determine the maximum allowable span of floor joists that support all other areas of the building, other than sleeping rooms and *attics*, provided that the design live load does not exceed 40 pounds per square foot (1.92 kPa) and the design dead load does not exceed 20 pounds per square foot (0.96 kPa).

TABLE R502.3.1(2)
FLOOR JOIST SPANS FOR COMMON LUMBER SPECIES
(Residential living areas, live load = 40 psf, L/D = 360)[b]

JOIST SPACING (INCHES)	SPECIES AND GRADE		DEAD LOAD=10 psf				DEAD LOAD=20 psf			
			2x6	2x8	2x10	2x12	2x6	2x8	2x10	2x12
			Maximum floor joist spans							
			(ft - in.)	(ft - in.)	(ft - in.)	(ft - in.)	(ft - in.)	(ft - in.)	(ft - in.)	(ft - in.)
12	Douglas fir-larch	SS	11-4	15-0	19-1	23-3	11-4	15-0	19-1	23-3
	Douglas fir-larch	#1	10-11	14-5	18-5	22-0	10-11	14-2	17-4	20-1
	Douglas fir-larch	#2	10-9	14-2	17-9	20-7	10-6	13-3	16-3	18-10
	Douglas fir-larch	#3	8-8	11-0	13-5	15-7	7-11	10-0	12-3	14-3
	Hem-fir	SS	10-9	14-2	18-0	21-11	10-9	14-2	18-0	21-11
	Hem-fir	#1	10-6	13-10	17-8	21-6	10-6	13-10	16-11	19-7
	Hem-fir	#2	10-0	13-2	16-10	20-4	10-0	13-1	16-0	18-6
	Hem-fir	#3	8-8	11-0	13-5	15-7	7-11	10-0	12-3	14-3
	Southern pine	SS	11-2	14-8	18-9	22-10	11-2	14-8	18-9	22-10
	Southern pine	#1	10-11	14-5	18-5	22-5	10-11	14-5	18-5	22-5
	Southern pine	#2	10-9	14-2	18-0	21-9	10-9	14-2	16-11	19-10
	Southern pine	#3	9-4	11-11	14-0	16-8	8-6	10-10	12-10	15-3
	Spruce-pine-fir	SS	10-6	13-10	17-8	21-6	10-6	13-10	17-8	21-6
	Spruce-pine-fir	#1	10-3	13-6	17-3	20-7	10-3	13-3	16-3	18-10
	Spruce-pine-fir	#2	10-3	13-6	17-3	20-7	10-3	13-3	16-3	18-10
	Spruce-pine-fir	#3	8-8	11-0	13-5	15-7	7-11	10-0	12-3	14-3
16	Douglas fir-larch	SS	10-4	13-7	17-4	21-1	10-4	13-7	17-4	21-0
	Douglas fir-larch	#1	9-11	13-1	16-5	19-1	9-8	12-4	15-0	17-5
	Douglas fir-larch	#2	9-9	12-7	15-5	17-10	9-1	11-6	14-1	16-3
	Douglas fir-larch	#3	7-6	9-6	11-8	13-6	6-10	8-8	10-7	12-4
	Hem-fir	SS	9-9	12-10	16-5	19-11	9-9	12-10	16-5	19-11
	Hem-fir	#1	9-6	12-7	16-0	18-7	9-6	12-0	14-8	17-0
	Hem-fir	#2	9-1	12-0	15-2	17-7	8-11	11-4	13-10	16-1
	Hem-fir	#3	7-6	9-6	11-8	13-6	6-10	8-8	10-7	12-4
	Southern pine	SS	10-2	13-4	17-0	20-9	10-2	13-4	17-0	20-9
	Southern pine	#1	9-11	13-1	16-9	20-4	9-11	13-1	16-4	19-6
	Southern pine	#2	9-9	12-10	16-1	18-10	9-6	12-4	14-8	17-2
	Southern pine	#3	8-1	10-3	12-2	14-6	7-4	9-5	11-1	13-2
	Spruce-pine-fir	SS	9-6	12-7	16-0	19-6	9-6	12-7	16-0	19-6
	Spruce-pine-fir	#1	9-4	12-3	15-5	17-10	9-1	11-6	14-1	16-3
	Spruce-pine-fir	#2	9-4	12-3	15-5	17-10	9-1	11-6	14-1	16-3
	Spruce-pine-fir	#3	7-6	9-6	11-8	13-6	6-10	8-8	10-7	12-4
19.2	Douglas fir-larch	SS	9-8	12-10	16-4	19-10	9-8	12-10	16-4	19-2
	Douglas fir-larch	#1	9-4	12-4	15-0	17-5	8-10	11-3	13-8	15-11
	Douglas fir-larch	#2	9-1	11-6	14-1	16-3	8-3	10-6	12-10	14-10
	Douglas fir-larch	#3	6-10	8-8	10-7	12-4	6-3	7-11	9-8	11-3
	Hem-fir	SS	9-2	12-1	15-5	18-9	9-2	12-1	15-5	18-9
	Hem-fir	#1	9-0	11-10	14-8	17-0	8-8	10-11	13-4	15-6
	Hem-fir	#2	8-7	11-3	13-10	16-1	8-2	10-4	12-8	14-8
	Hem-fir	#3	6-10	8-8	10-7	12-4	6-3	7-11	9-8	11-3
	Southern pine	SS	9-6	12-7	16-0	19-6	9-6	12-7	16-0	19-6
	Southern pine	#1	9-4	12-4	15-9	19-2	9-4	12-4	14-11	17-9
	Southern pine	#2	9-2	12-1	14-8	17-2	8-8	11-3	13-5	15-8
	Southern pine	#3	7-4	9-5	11-1	13-2	6-9	8-7	10-1	12-1
	Spruce-pine-fir	SS	9-0	11-10	15-1	18-4	9-0	11-10	15-1	17-9
	Spruce-pine-fir	#	8-9	11-6	14-1	16-3	8-3	10-6	12-10	14-10
	Spruce-pine-fir	#2	8-9	11-6	14-1	16-3	8-3	10-6	12-10	14-10
	Spruce-pine-fir	#3	6-10	8-8	10-7	12-4	6-3	7-11	9-8	11-3
24	Douglas fir-larch	SS	9-0	11-11	15-2	18-5	9-0	11-11	14-9	17-1
	Douglas fir-larch	# 1	8-8	11-0	13-5	15-7	7-11	10-0	12-3	14-3
	Douglas fir-larch	#2	8-1	10-3	12-7	14-7	7-5	9-5	11-6	13-4
	Douglas fir-larch	#3	6-2	7-9	9-6	11-0	5-7	7-1	8-8	10-1
	Hem-fir	SS	8-6	11-3	14-4	17-5	8-6	11-3	14-4	16-10[a]
	Hem-fir	#1	8-4	10-9	13-1	15-2	7-9	9-9	11-11	13-10
	Hem-fir	#2	7-11	10-2	12-5	14-4	7-4	9-3	11-4	13-1
	Hem-fir	#3	6-2	7-9	9-6	11-0	5-7	7-1	8-8	10-1
	Southern pine	SS	8-10	11-8	14-11	18-1	8-10	11-8	14-11	18-1
	Southern pine	#1	8-8	11-5	14-7	17-5	8-8	11-3	13-4	15-11
	Southern pine	#2	8-6	11-0	13-1	15-5	7-9	10-0	12-0	14-0
	Southern pine	#3	6-7	8-5	9-11	11-10	6-0	7-8	9-1	10-9
	Spruce-pine-fir	SS	8-4	11-0	14-0	17-0	8-4	11-0	13-8	15-11
	Spruce-pine-fir	#1	8-1	10-3	12-7	14-7	7-5	9-5	11-6	13-4
	Spruce-pine-fir	#2	8-1	10-3	12-7	14-7	7-5	9-5	11-6	13-4
	Spruce-pine-fir	#3	6-2	7-9	9-6	11-0	5-7	7-1	8-8	10-1

For SI: 1 inch = 25.4 mm, 1 foot = 304.8 mm, 1 pound per square foot = 0.0479 kPa.

Note: Check sources for availability of lumber in lengths greater than 20 feet.

a. End bearing length shall be increased to 2 inches.

b. Dead load limits for townhouses in Seismic Design Category C and all structures in Seismic Design Categories D_0, D_1, and D_2 shall be determined in accordance with Section R301.2.2.2.1.

Discussion: The IRC provides two different floor joist span tables, one of which is specifically for the reduced live load design requirement of 30 psf (1.44 kPa) that can only be used in sleeping rooms. Unfortunately, installing a hammock on a deck will not make it a sleeping room, so you must use the table based on the full 40 psf (1.92 kPa) design live-load requirement for all deck portions of the structure. Two columns are given, depending on the weight of the material itself, the dead load. In most cases, the dead loads of a typical deck will fall under the 10 psf (0.479 kPa) column. Heavy decking material, such as stone or concrete, may exceed a cumulative weight of materials of 10 psf (0.479 kPa), and thereby require you to use the reduced spans under the 20 psf (0.958 kPa) dead load column. While these floor span tables are meant to be generic, they do not account for many span reductions that must be considered for typical deck construction. Wet-use locations, material incising, chemical treatment and concentrated loads are not accounted for in these tables, yet may be present in deck construction. However, a direct reference to AF&PA span tables is provided in this section. Similar to the *Wood Frame Construction Manual*, the AF&PA span tables are a referenced standard, not an alternative. Use of AF&PA span tables can be done without specific approval as an "alternative," provided they fully apply to the criteria of the proposed design. These span tables are available as a free download from the American Wood Council web site, www.awc.org (see Appendix).

When using any span table, careful attention must be given to the species and grade of the lumber. As discussed earlier in this chapter under Section R502.1, lumber must be provided with a grade marking. This marking will provide the information necessary to properly determine the allowable span of various materials.

R502.3.3 Floor cantilevers. Floor cantilever spans shall not exceed the nominal depth of the wood floor joist. Floor cantilevers constructed in accordance with Table R502.3.3(1) shall be permitted when supporting a light-frame bearing wall and roof only. Floor cantilevers supporting an exterior balcony are permitted to be constructed in accordance with Table R502.3.3(2).

TABLE R502.3.3(2)
CANTILEVER SPANS FOR FLOOR JOISTS SUPPORTING EXTERIOR BALCONY[a, b, e, f]

Member Size	Spacing	Maximum Cantilever Span (Uplift Force at Backspan Support in lb)[c, d]		
		Ground Snow Load		
		≤ 30 psf	50 psf	70 psf
2 x 8	12″	42″ (139)	39″ (156)	34″ (165)
2 x 8	16″	36″ (151)	34″ (171)	29″ (180)
2 x 10	12″	61″ (164)	57″ (189)	49″ (201)
2 x 10	16″	53″ (180)	49″ (208)	42″ (220)
2 x 10	24″	43″ (212)	40″ (241)	34″ (255)
2 x 12	16″	72″ (228)	67″ (260)	57″ (268)
2 x 12	24″	58″ (279)	54″ (319)	47″ (330)

For SI: 1 inch = 25.4 mm, 1 pound per square foot = 0.0479 kPa.

a. Spans are based on No. 2 Grade lumber of Douglas fir-larch, hem-fir, southern pine, and spruce-pine-fir for repetitive (3 or more) members.
b. Ratio of backspan to cantilever span shall be at least 2:1.
c. Connections capable of resisting the indicated uplift force shall be provided at the backspan support.
d. Uplift force is for a backspan to cantilever span ratio of 2:1. Tabulated uplift values are permitted to be reduced by multiplying by a factor equal to 2 divided by the actual backspan ratio provided (2/backspan ratio).
e. A full-depth rim joist shall be provided at the unsupported end of the cantilever joists. Solid blocking shall be provided at the supported end.
f. Linear interpolation shall be permitted for ground snow loads other than shown.

Discussion: The first sentence of this section is very conservative in stating that a cantilever's horizontal projection cannot exceed the depth of the floor joist, but this is only a "general" requirement, allowing any loading condition to occur at the end of the cantilever. However, this section references two tables for more specific cantilever conditions supporting reduced additional loads, and allowing for much greater cantilever projections.

Of the two tables referenced in this section, only the one related to cantilevered balconies has been provided and discussed in this book. Table R502.3.3(2) is specifically referring to exterior balcony joists, and has considered the effects of snow load and live load in the allowable cantilever projections (see Example 4-23). The only caveat in applying this table to a deck cantilever, rather than a balcony cantilever, is that the back span of the floor joists is not inside the building (as intended by this IRC section); rather, it is exterior and would have a slightly reduced structural capacity due to the wet-use environment. Exterior balconies in the previous editions of the IRC were required to support a 60 psf (2.874 kPa) live load, compared to decks which must resist a 40 psf (1.92 kPa) live load. The 2009 IRC now requires a 40 pound (178 N) live load for balconies and decks.

Note c in Table R502.3.3(2) explains that the values in parenthesis within the table are the uplift forces that must be resisted at the back-spanned end of the cantilevered joist, the end opposite the cantilever. Like a child's teeter-totter, when the cantilevered portion of the joist is fully loaded, the other end will have a tendency to rise. Most hardware manufacturers will provide maximum uplift values in the load tables for their products. The hardware used at this end must be verified that it can accommodate resistance to the values in the IRC table. Note d is also related to the stress created in the back-spanned portion of cantilevered joists from a loaded overhang. The length of the backspan must be at least equal to two times the length of the overhang. This provides a sufficient length of material to resist the uplift at the backspan and bending at the support created by a loaded cantilever overhang.

Another important note to address is note e. The downward force on the cantilevered joists can cause the joists to rotate to their side and compromise their structural performance. The joists must be provided a resistance to this rotation at the ends of the cantilevered joists at the overhang and at the bearing location just prior to the overhang.

There is another way to exceed the general maximum projection of the depth of the joists provided in the first sentence of Section R502.3.3. The *Wood Frame Construction Manual* (WFCM), referenced by Section R301.1.1, contains a different method of determining maximum allowable cantilevers. This document allows a joist cantilever projection to equal one-fourth of its backspan. This method is foolproof, as long as you are using the appropriate span table for the span of the joist between the two bearing locations. By basing the cantilever projection on the allowable span of the joist, which in turn is based on the properties of the joist, it is a direct proportional relationship (see Example 4-24).

In using any of these methods for determining maximum cantilever projections, only uniformly distributed loads can be applied. Hanging a stairway from the cantilevered projection of a deck joist would apply loads that have not been considered in any of these methods and may require an alternative means to generate the design. This would also be required for any other significant dead loads applied to the cantilevered portion, such as a large firepit or built-in BBQ kitchen.

Example 4-23: The large cantilever in this photo is within the parameters of Table 502.3.3(2). A similar cantilever could be provided in a deck where the backspan is not actually inside the structure, but rather part of the deck.

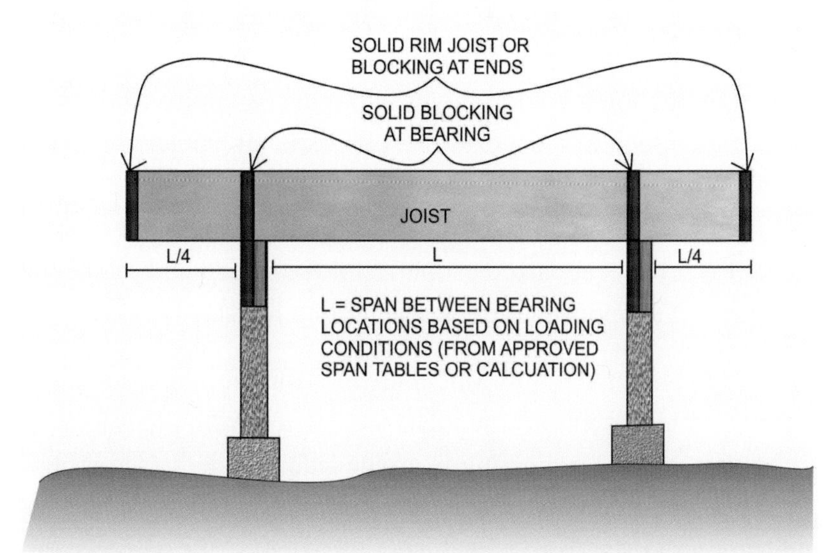

Example 4-24: The *Wood Frame Construction Manual* provides a simple method of determining a maximum allowable cantilever, as illustrated in this detail.

R502.10 Framing of openings. Openings in floor framing shall be framed with a header and trimmer joists. When the header joist span does not exceed 4 feet (1219 mm), the header joist may be a single member the same size as the floor joist. Single trimmer joists may be used to carry a single header joist that is located within 3 feet (914 mm) of the trimmer joist bearing. When the header joist span exceeds 4 feet (1219 mm), the trimmer joists and the header joist shall be doubled and of sufficient cross section to support the floor joists framing into the header. *Approved* hangers shall be used for the header joist to trimmer joist connections when the header joist span exceeds 6 feet (1829 mm). Tail joists over 12 feet (3658 mm) long shall be supported at the header by framing anchors or on ledger strips not less than 2 inches by 2 inches (51 mm by 51 mm).

Discussion: As has been previously discussed, the IRC floor framing provisions are designed with the intention of braced wall panels providing for lateral resistance—not something available for post-frame (post-and-beam) construction of exterior decks. In fundamentally applying this section to deck floor framing, the connection provisions are not applicable as they are primarily designed to resist vertical loads in a dry, controlled environment. These connections in post-frame deck construction are alternatives and must be approved by the building official, which in most cases results in metal hangers or framing anchors. What is still applicable in this section is the span and number of member plies required for carrying joists from other joists, provided the floor joists are sized properly for their span and load conditions and are not supporting other concentrated loads.

While this section is intended for openings in floors, it may also be used for other arrangements where joists are carried from other joists, such as in a stair landing projecting from the side of a deck (see Example 4-25). This section involves three types of joists—tail joists, header joists and trimmer joists; from a load path perspective, the trimmer carries the header which carries the tail joists. When the span of the header joist is 4 feet (1219 mm) or less, a single ply can be used to support the few tail joists that would attach to it. If the header joist is more than 4 feet (1219 mm) in length, it must be a double member. A trimmer joist that supports a single header joist can also be a single member, provided the header joist attaches and bears at the trimmer joist within 3 feet (914 mm) of the trimmer joist's bearing location. The trimmer joist must be a doubled member anytime the header joist exceeds 4 feet (1219 mm) in length or when the header joists does not bear within 3 feet (914 mm) of the trimmer joist bearing (see Example 4-26).

When using a joist to carry other joists, the supporting joist essentially becomes a beam. While the increased load and bending forces on the joist spans have been accounted for in this section, the concentrated load resulting at the end of the trimmer joist or header joist has not been accounted for in the ledger connection provisions in Table R502.2.2.1, previously discussed in this chapter. R502.2.2.2 specifically does not allow joists, beams or girders with increased loads, such as trimmer or header joists, to bear at a ledger.

Example 4-25: This photo of the underside of a stair landing illustrates how Section R502.10 can apply to deck floor arrangements where joists are supporting other joists. The beam connected to the ledger in this photo does not comply with Section R502.2.2.2.

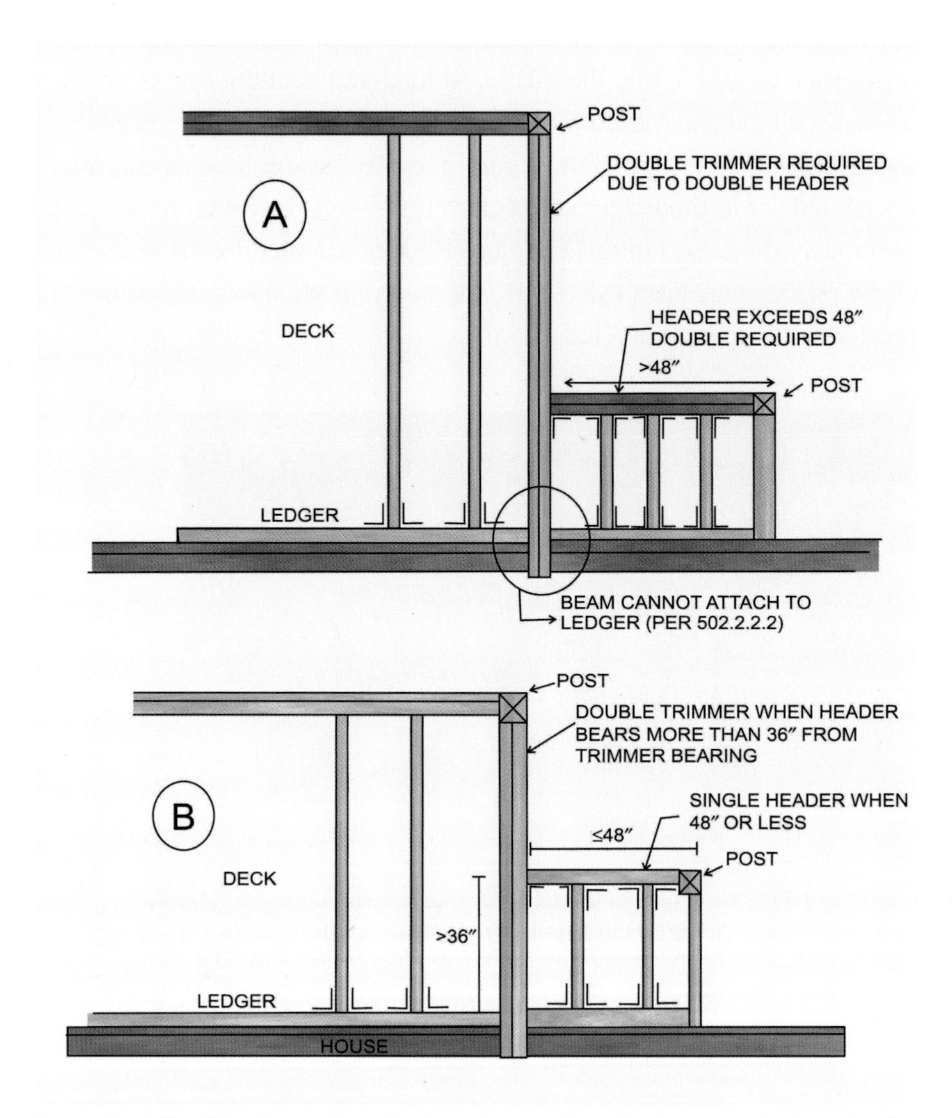

Example 4-26: This illustration displays the relationship between the length of the header joist, its bearing location on the trimmer joist and the number of joist plies required.

R502.6 Bearing. The ends of each joist, beam or girder shall have not less than 1.5 inches (38 mm) of bearing on wood or metal and not less than 3 inches (76 mm) on masonry or concrete except where supported on a 1-inch-by-4-inch (25.4 mm by 102 mm) ribbon strip and nailed to the adjacent stud or by the use of *approved* joist hangers.

Discussion: Three methods of bearing wood joists and beams are presented in this section. When a wood member is bearing directly on the resisting element, it is critical to provide a sufficient area of contact between the members. Loads imposed perpendicular to the grain of lumber deform the material much more than loads imposed parallel to the grain. If there is not enough bearing area provided in the contact of the loaded member and the resisting member, the loads are concentrated over too small an area, and the wood material may crush and settle. This may cause excessive deflection and damaged material, and may even cause the member to slip from its bearing location. Based on the common loads, spans and material sizes of prescriptive IRC design, at least $1^1/_2$-inches (38 mm) of bearing length must be provided for each joist and each member of a multi-ply beam. In the case of bearing on masonry or concrete, the shear strength of the concrete along the edge resisting the loads is considered, and thus the IRC requires at least 3 inches (76 mm) of bearing length. It is important to realize that all individual members of a multi-ply beam must be supported on both ends in accordance with this section (see Example 4-27)

The final method is the use of "approved" joist hangers. Most building officials will approve joist hangers provided the structural capacities provided by the manufacturer are not exceeded, the manufacturer corrosion-resistant recommendations are employed and the hanger is installed in accordance with the manufacturer's requirements (see Hardware discussion in Part 2 of this chapter).

Example 4-27: The beam ply to the left does not bear at least $1^1/_2$ inches (38 mm) on the post and does not comply with Section R502.6. All splices of beam plies must be supported in order to satisfy the prescriptive design provisions. For this project, a site-specific engineer evaluation provided evidence that the load path was sufficient as is, and the alternative was approved.

R502.6.2 Joist framing. Joists framing into the side of a wood girder shall be supported by *approved* framing anchors or on ledger strips not less than nominal 2 inches by 2 inches (51 mm by 51 mm).

Discussion: This section basically repeats the provisions in the previously discussed section, but does provide slightly more specific information in regard to joists framed into the side of a beam. Again, approved framing anchors (joist hangers) are required.

The method of using a ledger strip for deck joists' connections is not acceptable due to the lack of braced walls to resist the lateral forces from allowing the joists to slip off the ledger strip. This section, like most structural IRC provisions, cannot be used to its complete extent for deck construction due to the lack of post-frame (post-and-beam) provisions for lateral resistance.

R502.6.1 Floor systems. Joists framing from opposite sides over a bearing support shall lap a minimum of 3 inches (76 mm) and shall be nailed together with a minimum three 10d face nails. A wood or metal splice with strength equal to or greater than that provided by the nailed lap is permitted.

Discussion: If a center dropped beam is provided in a deck and joists are spliced at this beam, it is important for the joists to distribute their load evenly over the top of the beam. The live loads on the deck will not necessarily be evenly distributed on the joists at each side of the beam. By tying the joists together, they act as one unit as they bear on the top of the beam, thus maintaining the stability of the beam

Another important aspect of this splice connection is to resist lateral forces that tend to pull the deck apart at the beam. When uneven loading occurs on the deck, this connection completes the path of the lateral load resistance from the outer joists to the inner joists, the ledger and down to the foundation.

To provide this structural stability, the IRC requires joists to be lapped a minimum of 3 inches (76 mm), and fastened together with at least three 10d nails. The lap and the nailing will allow the uneven forces imposed on the joists to act as one force, concentrically loaded on the beam below. In deck construction, due to the exposed decking fasteners, this offset in joists is typically undesirable. In lieu of the lap and face nailing, the IRC allows a wood or metal splice to be used at the side of each joist and over the bearing location (see Example 4-28). This splice is required to be made of equal or greater strength than the lapped connection using the three 10d nails. Logically thinking, this would require at least three 10d nails through the splicing member into each joist, for a total of six nails.

Example 4-28: The metal plate used to tie this beam together is an example of how deck joists spliced at a bearing location can be tied together in accordance with Section R502.6.1. Though not specifically required for beams by the IRC, securing a beam splice together is a good practice of wood frame construction.

R502.7 Lateral restraint at supports. Joists shall be supported laterally at the ends by full-depth solid blocking not less than 2 inches (51 mm) nominal in thickness; or by attachment to a full-depth header, band or rim joist, or to an adjoining stud or shall be otherwise provided with lateral support to prevent rotation.

Exception:

1. Trusses, structural composite lumber, structural glued-laminated members and I-joists shall be supported laterally as required by the manufacturer's recommendations.
2. In Seismic Design Categories D_0, D_1 and D_2, lateral restraint shall also be provided at each intermediate support.

R502.7.1 Bridging. Joists exceeding a nominal 2 inches by 12 inches (51 mm by 305 mm) shall be supported laterally by solid blocking, diagonal bridging (wood or metal), or a continuous 1-inch-by-3-inch (25.4 mm by 76 mm) strip nailed across the bottom of joists perpendicular to joists at intervals not exceeding 8 feet (2438 mm).

Exception: Trusses, structural composite lumber, structural glued-laminated members and I-joists shall be supported laterally as required by the manufacturer's recommendations.

Discussion: When a joist is restrained in a vertical position, it performs as expected. If not braced in a vertical position, it may rotate to the side under loading conditions. This rotation will create more deflection in the deck surface and subject the newly oriented joists to conditions they were not designed for.

In mid-span, most joists may slightly twist without any real concern, as the decking attached to the top edge will transmit the rotational force through multiple members. In the case that nominal lumber greater than 2 inches by 12 inches (51 mm by 305 mm) is used as joists, a means to resist this rotation within the span at intervals not exceeding 8 feet (2438 mm) will be required.

At the bearing locations, however, a twisting or rotating joist may affect stability, bearing capacity, bearing area and connection to the beam. For this reason, the IRC requires all joists to be braced at their ends against rotation by lateral support provided by full depth blocking or end connection to a full-depth beam or rim/band joist, Although the title of the section is "restraint at supports," the language of the section requires joists to be laterally supported "at the ends." The real intent is that joists be restrained against rotation at all bearing points.

> **Side Note:** Blocking joists at mid-span will considerably stiffen the deck surface, as they resist the joist rotation discussed above. Keeping the joists vertical throughout their span will dramatically reduce deflection and bounce.

R502.8 Drilling and notching. Structural floor members shall not be cut, bored or notched in excess of the limitations specified in this section. See Figure R502.8.

R502.8.1 Sawn lumber. Notches in solid lumber joists, rafters and beams shall not exceed one-sixth of the depth of the member, shall not be longer than one-third of the depth of the member and shall not be located in the middle one-third of the span. Notches at the ends of the member shall not exceed one-fourth the depth of the member. The tension side of members 4 inches (102 mm) or greater in nominal thickness shall not be notched except at the ends of the members. The diameter of holes bored or cut into members shall not exceed one-third the depth of the member. Holes shall not be closer than 2 inches (51 mm) to the top or bottom of the member, or to any other hole located in the member. Where the member is also notched, the hole shall not be closer than 2 inches (51 mm) to the notch.

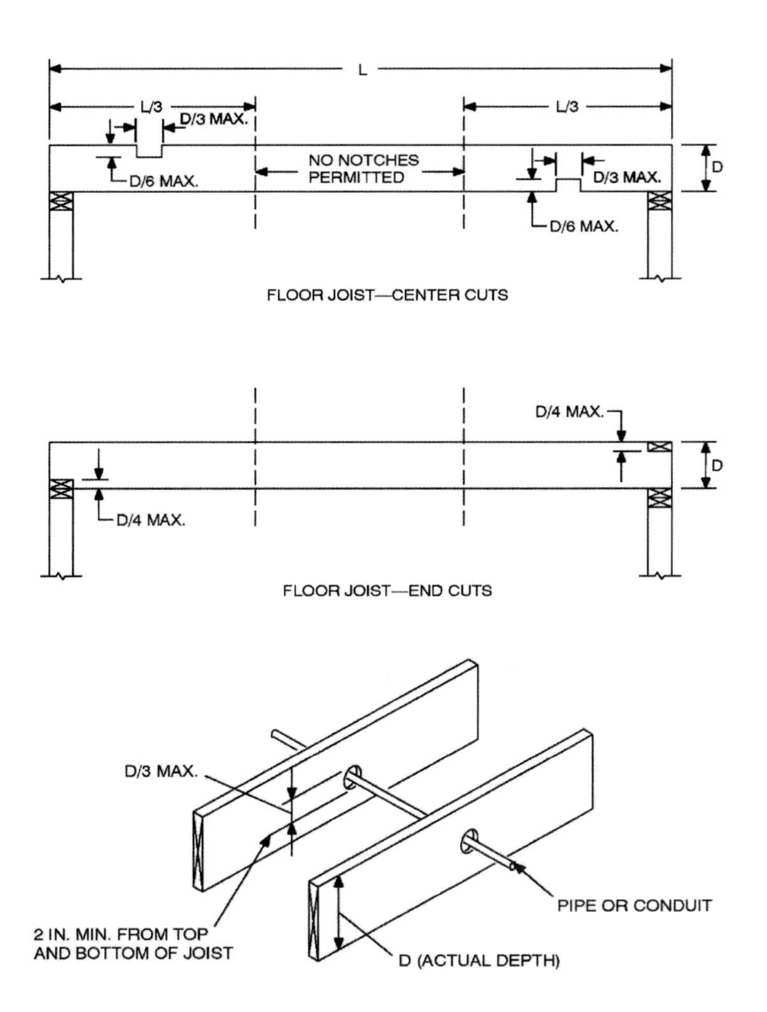

For SI: 1 inch = 25.4 mm.

IRC Figure R502.8: Cutting, Notching and Drilling

R502.8.2 Engineered wood products. Cuts, notches and holes bored in trusses, structural composite lumber, structural glue-laminated members or I-joists are prohibited except where permitted by the manufacturer's recommendations or where the effects of such alterations are specifically considered in the design of the member by a *registered design professional.*

Discussion: There are many internal forces acting within a joist which can be compromised by careless alterations to the material cross-section. When notching lumber, the top or bottom edges are compromised (the tension or compression fibers), and will thus decrease the members bending resistance. For this reason, notches are prohibited in the middle third of the span, where the bending stress is the greatest. In the remaining portions of the span, notches must be within the limits detailed in Figure R502.8. These limitations are based on the actual measured depth of the framing material, referred to as "D" in Figure R502.8. Allowable notching at the bearing points of up to one-fourth the depth of the member can allow for considerable design flexibility, without reducing the allowable span of the material (see Example 4-29). When using material with a nominal thickness of greater than 4 inches (102 mm), such as a 4 by 10 Doug-Fir timber only, the tension edge (the bottom edge) can be notched at the bearing locations.

Holes cannot be greater in diameter than one-third the member's actual depth "D," but also must maintain 2 vertical inches (51 mm) of unaltered material at the top and bottom of the member, so the tension and compression fibers are not compromised. A 2 by 6, for example, can be drilled to a maximum diameter of 1$^1/_2$ inches (38 mm) in order to maintain the 2 inch (51 mm) top and bottom requirement, even though this is less than one-third the member's depth. Holes also must be at least 2 inches (51 mm) from other holes or notches in the same member.

When working with any engineered wood products, these provisions are not applicable. The individual manufacturer of the wood product has the authority as to the allowable modifications of their product, both notching and drilling. These limitations are usually readily available by manufacturers and can vary greatly depending on the type of product. In the event that differing means of modifying these products is desired, a registered design professional is typically able to perform the necessary engineering calculations, provided the properties of the material used in the calculations are based on those published by the manufacturer.

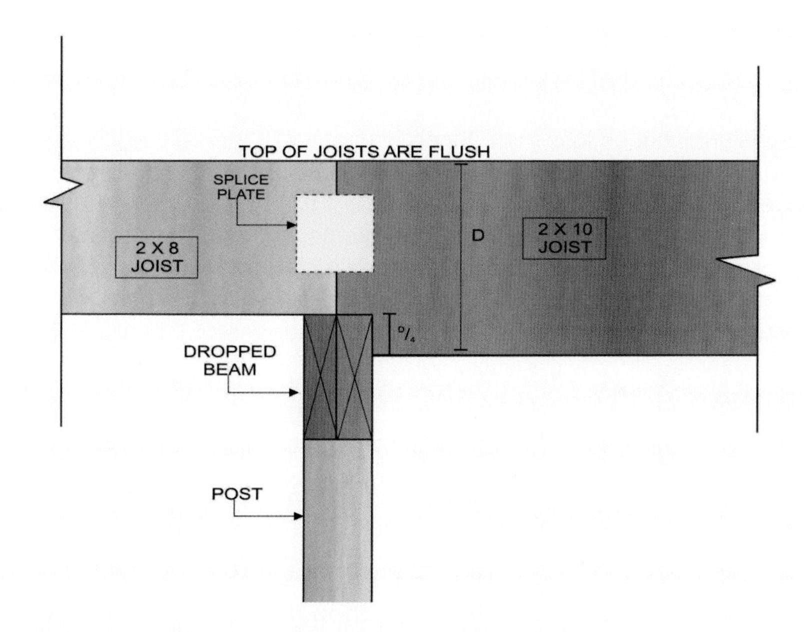

Example 4-29: Notching joists at the bearing location allows for an increased flexibility in the design and material choice. This example illustrates how a 2 by 8 and 2 by 10 joist can be altered such that the top edges are flush. A notch can also occur where a joist bears in a metal joist hanger.

R502.9 Fastening. Floor framing shall be nailed in accordance with Table R602.3(1). Where posts and beam or girder construction is used to support floor framing, positive connections shall be provided to ensure against uplift and lateral displacement.

Discussion: The fastener schedule referenced in the first sentence of Section R502.9 is not applicable to typical post-frame (post-and-beam) deck construction. What is applicable for deck construction is the clarification of the requirement for a complete load path to the foundation for all forces imposed, as specified in Section R301.1. Section R502.9 mentions post-and-beam construction, and makes a general comment that the connections supporting the floor system must be capable of resisting uplift and lateral displacement. However, there is no prescriptive connection to provide this in the fastener table, and most often the use of metal hardware, such as joist clips, and post caps will be required.

R407.3 Structural requirements. The columns shall be restrained to prevent lateral displacement at the bottom end. Wood columns shall not be less in nominal size than 4 inches by 4 inches (102 mm by 102 mm). Steel columns shall not be less than 3-inch-diameter (76 mm) Schedule 40 pipe manufactured in accordance with ASTM A 53 Grade B or *approved* equivalent.

> **Exception:** In Seismic Design Categories A, B and C, columns no more than 48 inches (1219 mm) in height on a pier or footing are exempt from the bottom end lateral displacement requirement within under-floor areas enclosed by a continuous foundation.

Discussion: The previous IRC Section R502.9 clarifies the load resistance required by the top of the post and the beam or joist connection, and Section R407.3 discusses the bottom of this post or "column," at its connection to the foundation (see Example 4-30). Again reinforcing the requirement for a complete load path, it is made clear that a post must resist lateral forces by its connection to the foundation system. The exception to this requirement is not applicable to deck construction.

Also in this section, a minimum size for wood and steel columns is provided. This minimum is only meant to require that no wood post can be less than a 4 by 4, not that all posts can be as small as a 4 by 4. In platform or balloon framing, the sheathing attached to wall studs provides lateral resistance against the buckling or bending of the studs. In post-frame (post-and-beam) construction, however, the cross-section of the post must be capable of resisting bending forces on its own or by the use of bracing. Depending on the vertical load imposed on the post, the height and species of the post and the use of additional bracing, a post may be required to be larger than 4 inches by 4 inches (102 mm by 102 mm) in order to safely transmit all loads to the foundation. At this time in IRC development there are no prescriptive design parameters for properly sizing and designing wood posts for post-frame construction. However, there are alternative sources for this information. For example, the Forest Products Society 1996 book title, *Wood Decks*, provides a post sizing table developed by AF&PA which specifies a range of maximum 4 by 4 post heights from 4 feet (1219 mm) when supporting 256 square feet (23.8 m^2) of deck to a maximum height of 10 feet (3048 mm) when supporting only 36 square feet (3.3 m^2) of deck. This type of information from reputable sources can be submitted as an alternative design (see Appendix).

Example 4-30: The use of a post base will secure the base of a post against lateral forces. If the post (pressure-preservative-treated for ground contact use) was sunk into the earth and bearing on a footing below, the soil surrounding the post will be sufficient to resist these lateral forces.

Chapter 5: Decking

Introduction

Exterior decking material has historically been that of lumber, varying in species from coast to coast. The east coast has primarily employed preservative-treated Southern Yellow Pine or Fir as the decking of choice, while the west coast has favored Redwood and Cedar. However, in the past decade, a major boom has occurred in the decking industry, providing for a much larger variety of decking materials, textures, colors, shapes, fastening methods, and of course, final cost to the consumer. There are more than 30 different decking manufacturers in the United States, most of which boast a unique product through variations of the characteristics mentioned above. Wood/plastic composite decking makes up a major percent of the manufactured decking market, yet can vary considerably in composition. The wood species, type of plastic, recycled or virgin material, and the technology used to put it together can create dramatically unique finished products. While composite decking is a readily available decking type, other manufactured materials are also available for decking application, such as vinyl, PVC, plastic, stone, tile, steel, aluminum, fiberglass and concrete.

Also expanding the market has been the use of hardwood lumber. While redwood and cedar remain the most common natural lumber decking, tropical hardwoods have also made their mark in many backyards. These hardwoods seem to be available in nearly as many colors, textures and finishes as the manufactured decking market. Ipe, Brazilian Redwood, Tigerwood, Garapa and Cumara are just a few of the other wood decking products available. The concern and difficulty in using these products is there is no grading evaluation available that addresses structural issues, as required by Section R502.1, discussed in Chapter 4 of this book. Unlike softwood species, where the grade stamp addresses strength values, the grade stamps on hardwood species only addresses the finish, but no structural considerations. Therefore, there is no means for an inspector to verify that the hardwood lumber chosen for the decking material will meet the provisions of the code. When considering the use of hardwood lumber for exterior decking, the local building official should be consulted.

With so many options of decking, all with specific installation requirements and structural capacities, it is impractical for there to be governing provisions in the IRC for all of them. As will be discovered throughout this chapter, there are very few decking-related IRC sections available. Many of the currently available decking products are considered "alternatives" (see Chapter 1).

Part One: Manufactured Decking

Definitions

WOOD/PLASTIC COMPOSITE (IRC). A composite material made primarily from wood or cellulose-based materials and plastic.

Discussion: A common misconception in construction is that if a product is manufactured, marketed and distributed across the United States, then it is acceptable for use throughout the United States. While for many products and applications this may be true, it is usually the result of extensive work that occurred long before mass distribution. When a material is not specifically addressed by the IRC, it is considered an alternative material and must be approved by the local building official (see Chapter 1 for more details). In approving alternative materials for use in their jurisdiction, a building official will request evidence that the material will perform in a manner equivalent to what is provided or required by the IRC. Most decking manufacturers provide this evidence in the form of test results rather than engineering, which must be based on known properties of the material composition. Manufactured decking is usually composed of a custom chemical blend of various materials, thus requiring a physical test to uncover its properties and capacities (see the discussion of Section R104.11.1 in Chapter 1). New provisions in the 2009 version of the IRC include a referenced standard for all wood and plastic composite decking, stair tread and guard installations, and are provided following this discussion.

R311.7.4.4 Exterior wood/plastic composite stair treads. Wood/plastic composite stair treads shall comply with the provisions of Section R317.4.

R312.4 Exterior woodplastic composite guards. Woodplastic composite *guards* shall comply with the provisions of Section R317.4.

R317.4 Wood/plastic composites. Wood/plastic composites used in exterior deck boards, stair treads, handrails and guardrail systems shall bear a *label* indicating the required performance levels and demonstrating compliance with the provisions of ASTM D 7032.

R317.4.1 Wood/plastic composites shall be installed in accordance with the manufacturer's instructions.

Discussion: Wood and plastic composite decking, stair treads and guard components must be tested under the criteria detailed in ASTM D 7032. This new referenced standard in the 2009 IRC allows wood and plastic composite decking to be installed without specific approval as an "alternative." Composite decking and other types of manufactured decking that have not been tested in accordance with this standard can still be installed through other test procedures and the acceptance by a local building official as an "alternative" material. When decking is tested to this standard, or approved as an alternative, all decking and stair treads must be installed in accordance with the manufacturer's written installation instructions, which most likely are based on the extensive testing of the product.

Many various limitations and installation requirements are provided in the manufacturer's installation instructions and test reports for the numerous variations and compositions of manufactured decking products. Examples of some of these requirements and limitations are provided in Examples 5-1, 5-2, 5-3 and 5-4.

The ICC Evaluation Service (ICC-ES), a subsidiary of the International Code Council, is a public-benefit corporation that does technical evaluations of building products, components, methods and materials. If it is found that the subject of an evaluation complies with code requirements, then ICC-ES publishes a report to that effect, and makes the report available to the public. These reports are intended to aid agencies that enforce building regulations by assisting them to determine code compliance, and companies that hold evaluation reports will find their products recognized by the public as code-complying by the ICC-ES.

ICC-ES has developed Acceptance Criteria which describe the standards and methodology used to test various composite decking products, referred to as *Acceptance Criteria for Deck Board Span Ratings and Guardrail Systems* (AC 174). Characteristics such as deflection, structural resistance to loading, temperature and moisture effects, ultraviolet resistance, freeze-thaw resistance, termite and decay resistance, flame spread and mechanical fastening methods, are all evaluated under specific criteria required by ASTM D 7032. The results of these various tests may create considerable variations in the performance capabilities of different manufactured decking products, thus reinforcing the importance of referring to each product's manufacturer installation instructions.

Example 5-1: The joists in this photo are spaced at 8 inches (203 mm) on center. This particular decking product, when installed on a 45 degree (0.79 rad) angle to the joists, requires a maximum spacing of 12 inches (305 mm) on center, as noted in the manufacturer's installation instructions. Originally installed on joists that were spaced 16 inches (406 mm) on center, additional joists were required to be installed such that the spacing did not exceed the 12 inch (305 mm) maximum.

Example 5-2: Wood and plastic composite decking undergoes expansion and contraction under temperature variations. Most decking products, including those other than wood/plastic composites will require evaluation of the installation temperatures and appropriate gapping, as required by the manufacturer. The deck boards in this example were installed with no gap during cold temperatures and expanded considerably when warm weather arrived, thus dramatically affecting the performance and stability of the installation.

Example 5-3: Manufactured decking products have limitations in their ability to extend beyond the edge of a support. Depending on the composition of the product, allowable overhangs may range anywhere from 0 to up to 4 inches (102 mm). The product in this photo was limited to a maximum 2-inch (51 mm) overhang. Stepping on the edge of this deck could result in failure of the decking material and a dangerous and unexpected fall off the edge.

Example 5-4: Decking products will usually require end bearing between $^3/_4$ of an inch to $1^1/_2$ inches (19.1 mm to 38 mm), depending usually on the magnitude of the possible expansion and contraction. The deck board to the left of this joint is installed with less than $^1/_2$ inch (12.7 mm) bearing on the double joist intended to provide $1^1/_2$ inches (38 mm) of bearing for each side of the joint. Under extremely cold temperatures, with maximum contraction of the decking material, this deck board could be pulled off the supporting joist entirely.

Part Two: Lumber Decking

R503.1 Lumber sheathing. Maximum allowable spans for lumber used as floor sheathing shall conform to Tables R503.1, R503.2.1.1(1) and R503.2.1.1(2).

R503.1.1 End joints. End joints in lumber used as subflooring shall occur over supports unless end-matched lumber is used, in which case each piece shall bear on at least two joists. Subflooring may be omitted when joist spacing does not exceed 16 inches (406 mm) and a 1-inch (25.4 mm) nominal tongue-and-groove wood strip flooring is applied perpendicular to the joists.

TABLE R503.1
MINIMUM THICKNESS OF LUMBER FLOOR SHEATHING

JOIST OR BEAM SPACING (INCHES)	MINIMUM NET THICKNESS	
	Perpendicular to Joist	Diagonal to Joist
24	11/16	3/4
16	5/8	5/8
48[a]	1 1/2 T & G	N/A
54[b]		
60[c]		

For SI: 1 inch = 25.4 mm, 1 pound per square inch = 6.895 kPa.
a. For this support spacing, lumber sheathing shall have a minimum F_b of 675 and minimum E of 1,100,000 (see AF&PA/NDS).
b. For this support spacing, lumber sheathing shall have a minimum F_b of 765 and minimum E of 1,400,000 (see AF&PA/NDS).
c. For this support spacing, lumber sheathing shall have a minimum F_b of 855 and minimum E of 1,700,000 (see AF&PA/NDS).

Discussion: The above listed IRC sections are intended to apply to interior subfloors of conventional construction utilizing dimensional lumber. For exterior deck construction, only parts of these sections are applicable. Although intuitive to most construction professionals, this section clarifies that each deck board end-to-end joint must occur over a supporting joist member. There is an exception for end-matched lumber, such as a tongue-and-groove profile, but this type of lumber material is not as commonplace in deck construction.

Table 503.1 provides the minimum thickness of lumber decking required for common joist spacing. Again, only parts of this table are applicable in deck construction. The 16-inch (406 mm) and 24-inch

(610 mm) joist spacing columns provide lumber thicknesses for orientation both perpendicular to the joists and diagonal to the joists. However, most lumber deck materials do not have a thickness less than a nominal $^{7}/_{8}$ inch (22.2 mm), generally making the minimums provided in this table well below the industry standard. Nominal lumber refers to the size of the material prior to finishing of the surface. In decking, a nominal 2-inch (51 mm) material is usually a measured 1$^{1}/_{2}$ inches (38 mm), also called "dimensional," "net" or "minimum dressed" thickness. Similarly, a 1$^{1}/_{4}$ inch (32 mm) nominal measures at 1 inch (25 mm) and a 1 inch (25 mm) nominal measures in at $^{3}/_{4}$ inch (19.1 mm).

While these sections are somewhat limited in their scope for exterior deck construction, they generally cover the typical application of lumber decking. Other installations, such as greater joist spacing, or use as stair treads, would be considered an alternative.

Fastening of lumber decking to supporting joists and beams is detailed in Table R602.3(1). As with other IRC provisions, the fastening criteria provided in this table for lumber decking is primarily intended for applications of interior subfloors, and is therefore not provided in this book. One-inch and 2-inch nominal lumber subflooring is listed in this table and requires 8d or 16d nails, respectively, as the fastening means. It is commonly accepted for #8 or greater wood screws to be used for exterior decking fastening, as the freeze/thaw and expansion/contraction of moisture and materials causes nails to rise out from the decking material. Hidden fastener systems and other types of fastening products must be tested as an alternative, and provide results equivalent to the nailing provided in the IRC. What is directly applicable to decking fastening in this table is the number of fasteners per board, per joist. In all lumber decking applications, two fasteners are required at each joist. Just as for other connections involving preservative-treated lumber, fasteners for decking secured to treated lumber must be in accordance with Section R317.3.1, discussed in Chapter 4 of this book.

Chapter 6: Stairways and Ramps

Introduction

It's a rare occurrence to find a deck that does not contain a change of elevation. Be it a step down from a door to the deck or a step out to the yard from the deck, stairways and sometimes ramps will most likely be present. Any time people traverse over and across sloping or uneven surfaces, or ascend or descend a stairway, there will be a certain level of instability and hazard. To reduce this hazard, the IRC provides very detailed and specific requirements, particularly in stairways.

When a person ascends or descends a stairway, his or her mind quickly recognizes the "rhythm" created by the person's legs and body. After only a couple steps, he or she will subconsciously begin to expect the next step to be in about the same spot as the previous. A sudden change in that rhythm, such as a variation in the rise, run or shape of a step, is the cause of many stair related injuries. Stair geometry can seem complicated, and once the stairway components are cut and constructed incorrectly, it is not easy to correct without a complete rebuild. Exercising patience when laying out the stair stringers can be worth its weight in gold when it comes to the final geometry of the finished stairway.

When reviewing the IRC requirements related to stairways, such as handrails, safety glazing and landings, it is very important to understand and distinguish a stair from a stairway, as many provisions depend on this distinction.

Definitions (see Examples 6-1, 6-2 and 6-3)

FLIGHT. (IRC) A continuous run of rectangular treads or winders or combination thereof from one landing to another.

NOSING. (IRC) The leading edge of treads of stairs and of landings at the top of stairway flights.

RAMP. (IRC) A walking surface that has a running slope steeper than 1 unit vertical in 20 units horizontal (5-percent slope).

STAIR. (IRC) A change in elevation, consisting of one or more risers.

STAIRWAY. (IRC) One or more flights of stairs, either exterior or interior, with the necessary landings and platforms connecting them to form a continuous and uninterrupted passage from one level to another within or attached to a building, porch or deck.

STAIRWAY, SPIRAL. (IBC) A *stairway* having a closed circular form in its plan view with uniform section-shaped treads attached to and radiating from a minimum-diameter supporting column.

WINDER. (IRC) A tread with nonparallel edges.

Example 6-1: The single step between these two deck levels is a stair and a stairway for the purpose of applying the provisions of the IRC.

Example 6-2: This single stairway, composed of three flights of stairs, two intermediate landings and a top and bottom landing, connects two "levels" of the deck and the patio below by an uninterrupted path.

Example 6-3: Two connected stairways are pictured here. The upper stair and the top and intermediate landing create an uninterrupted path from the deck to the garage door. A second stairway consists of the lower stair and the intermediate and bottom landing, and creates a path from the garage door to grade. This arrangement is not a single stairway because the path from the deck to grade is interrupted by the garage-door access at the intermediate landing. This distinction is important when analyzing the safety glazing requirements discussed in Chapter 2.

Part One: Stairways, General

R311.5 Construction.

R311.5.1 Attachment. Exterior landings, decks, balconies, stairs and similar facilities shall be positively anchored to the primary structure to resist both vertical and lateral forces or shall be designed to be self-supporting. Attachment shall not be accomplished by use of toenails or nails subject to withdrawal.

R301.5 Live load. The minimum uniformly distributed live load shall be as provided in Table R301.5.

TABLE R301.5
MINIMUM UNIFORMLY DISTRIBUTED LIVE LOADS
(in pounds per square foot)

USE	LIVE LOAD
Attics without storage[b]	10
Attics with limited storage[b,g]	20
Habitable attics and attics served with fixed stairs	30
Balconies (exterior) and decks[e]	40
Fire escapes	40
Guardrails and handrails[d]	200[h]
Guardrail in-fill components[f]	50[h]
Passenger vehicle garages[a]	50[a]
Rooms other than sleeping room	40
Sleeping rooms	30
Stairs	40[c]

For SI: 1 pound per square foot = 0.0479 kPa, 1 square inch = 645 mm^2,
1 pound = 4.45 N.

a. Elevated garage floors shall be capable of supporting a 2,000-pound load applied over a 20-square-inch area.

b. Attics without storage are those where the maximum clear height between joist and rafter is less than 42 inches, or where there are not two or more adjacent trusses with the same web configuration capable of containing a rectangle 42 inches high by 2 feet wide, or greater, located within the plane of the truss. For attics without storage, this live load need not be assumed to act concurrently with any other live load requirements.

c. Individual stair treads shall be designed for the uniformly distributed live load or a 300-pound concentrated load acting over an area of 4 square inches, whichever produces the greater stresses.

d. A single concentrated load applied in any direction at any point along the top.

e. See Section R502.2.2 for decks attached to exterior walls.

f. Guard in-fill components (all those except the handrail), balusters and panel fillers shall be designed to withstand a horizontally applied normal load of 50 pounds on an area equal to 1 square foot. This load need not be assumed to act concurrently with any other live load requirement.

g. For attics with limited storage and constructed with trusses, this live load need be applied only to those portions of the bottom chord where there are two or more adjacent trusses with the same web configuration capable of containing a rectangle 42 inches high or greater by 2 feet wide or greater, located within the plane of the truss. The rectangle shall fit between the top of the bottom chord and the bottom of any other truss member, provided that each of the following criteria is met.

1. The attic area is accessible by a pull-down stairway or framed in accordance with Section R807.1.
2. The truss has a bottom chord pitch less than 2:12.
3. Required insulation depth is less than the bottom chord member depth.

The bottom chords of trusses meeting the above criteria for limited storage shall be designed for the greater of the actual imposed dead load or 10 psf, uniformly distributed over the entire span.

h. Glazing used in handrail assemblies and guards shall be designed with a safety factor of 4. The safety factor shall be applied to each of the concentrated loads applied to the top of the rail, and to the load on the in-fill components. These loads shall be determined independent of one another, and loads are assumed not to occur with any other live load.

Discussion: Stairs can be constructed of many different materials and with many different methods, all of which could satisfy the geometric and structural requirements of the IRC. Even in the category of common deck stairs, there are more methods used across the United States than could be feasibly included in a model code. The IRC does not provide any pre-engineered, prescriptive method for the structural design of stair construction, but it does provide parameters that must be satisfied by the design. It has become commonplace in the construction of new homes to see interior premanufactured stairways engineered, constructed and delivered to the site as one complete system. Field-fabricated stairs, however, vary greatly in the design, workmanship and final product, and this is the type of stair construction typically seen with decks. Approving these stairways as sufficient for resisting the prescribed live load criteria requires the use of alternative designs from engineering sources or the knowledge and experience of the local building official in regard to historically accepted stair design. In all of the various methods of stair construction, it is important to consider both the capabilities of the stair tread and the stair stringers, and how they function together (see Example 6-4).

The attachment of a flight of stairs to a deck must be accomplished in a manner that will resist forces in both the vertical and horizontal directions. Deck stairs cannot be attached with toenails or with nails installed in a manner where they may be subject to withdrawal. Nails are considered "subject to withdrawal" when they are installed such that their load resistance capabilities are in a direction parallel to the shank of the nail. Nails are primarily designed and intended to be oriented so that they resist loads utilizing the shear strength of the nail—perpendicular to the shank.

As for most other structural components, the IRC provides minimum requirements for live load resistance of stairways and individual treads. Stairways as a whole must be capable of resisting a 40 pound per square foot (1.92 kPa) continuous and uniformly distributed live load. This live load resistance is intended to apply concurrently to all stair treads and landings that make up the stairway. Stair treads, as a component of the stairway system, have an additional and specific requirement in their load-resistant capability. Across the individual span of each tread, a 300-pound (1334 N) load concentrated over a 4-square-inch (2561 mm^2) area must be sufficiently resisted. In the testing and approval of manufactured composite decking materials, this concentrated load requirement often results in reduced span ratings of most products when used as stair treads (see Example 6-5). Many manufacturers have omitted this load in the testing of their product, and specifically refer the product user to the expertise of a registered design professional for use of their product in stair tread applications.

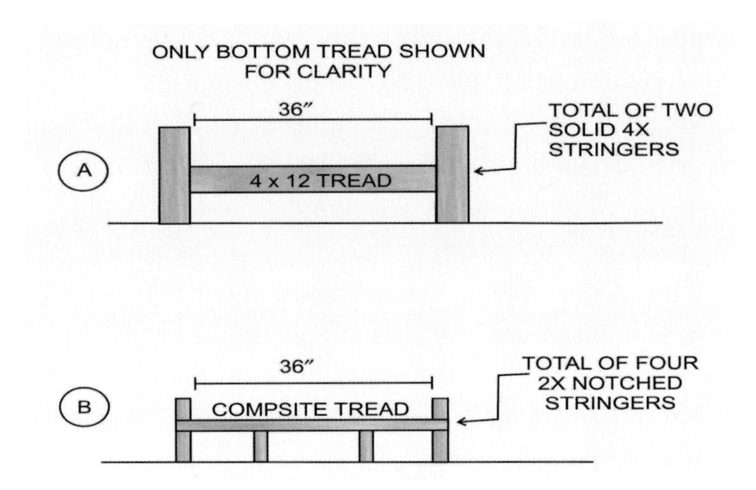

Example 6-4: In detail A, a very strong and sturdy tread material is used that is capable of spanning the full 36-inch (914 mm) width of the stairway; however, this design provides only two stringers that together must be capable of resisting a uniformly distributed 40 pound (178 N) live load applied to the entire stair. In detail B, a decking material is installed that requires a maximum 12-inch (305 mm) stringer spacing, resulting in four stringers to resist the same uniformly distributed 40 pound (178 N) live load as in detail A. The size and number of stringers required cannot be determined independently from the tread material used.

Example 6-5: When this heavy-timber stairway had its 4 by 12 treads replaced with a composite decking, the previous stringer spacing of 18 inches (457 mm) was insufficient based on the manufacturer's installation requirements to support the span of the new tread material, thus the addition of the two 2 by 12 notched stringers. A related issue, the thickness of the new treads, 1$^1/_4$ inch (32 mm) compared to the previous 4-inch (102 mm) thick treads, enlarged the size of the riser openings such that they no longer comply with the IRC; see the discussion of Section R311.7.4.3 in Part 2 of this chapter for riser opening limitations.

R311.7.1 Width. Stairways shall not be less than 36 inches (914 mm) in clear width at all points above the permitted handrail height and below the required headroom height. Handrails shall not project more than 4.5 inches (114 mm) on either side of the stairway and the minimum clear width of the stairway at and below the handrail height, including treads and landings, shall not be less than 31$^1/_2$ inches (787 mm) where a handrail is installed on one side and 27 inches (698 mm) where handrails are provided on both sides.

> **Exception:** The width of spiral stairways shall be in accordance with Section R311.7.9.1.

R311.7.2 Headroom. The minimum headroom in all parts of the stairway shall not be less than 6 feet 8 inches (2032 mm) measured vertically from the sloped line adjoining the tread nosing or from the floor surface of the landing or platform on that portion of the stairway.

Discussion: There is often a general expectation in deck construction that stairways must be at least 36 inches (914 mm) wide. While this width would certainly provide a nice comfortable stairway, one that is easy to move your patio furniture up, it is not the minimum requirement. A width of 36 inches (914 mm) is required, but not always at foot level. Generally speaking, our upper bodies are wider than our lower bodies. We need room for our shoulders and arms to swing around and carry items as we walk, but our legs are usually narrower and move in a straight line. This has been considered in the IRC and is reflected in the minimum requirements.

To understand where the 36-inch (914 mm) width is required, you must first understand some other height requirements of stairways. A vertical clearance (headroom) of 6 feet 8 inches (2032 mm) is required at all portions of a stairway, to include any landings between stairs, and at the top and bottom of the stairway (see "Stairway" definition). This headroom is measured from the nosing of each tread and from the surface of the landings (see Example 6-6). The headroom area must be at least 36 inches (914 mm) wide (measured parallel to the stair treads), but only in the region above any provided handrails and up to the minimum 6 feet 8 inches (2032 mm). It is important to remember that nothing can project into this required clearance, including light fixtures. Below the handrails, the 36-inch (914 mm) clearance is not always required—it depends on the existence of handrails.

If the stairway does not require handrails (discussed in Chapter 7), and there are no handrails installed, the 36-inch (914 mm) clearance is required for the entire height of the stairway, all the way from the treads. However, if there is a single handrail on one side, the handrail is allowed to project into the 36-inch width (914 mm) by up to $4^1/_2$ inches (114 mm), thus reducing the required width at and beneath the handrails to $31^1/_2$ inches (787 mm). In deck construction, handrails are often built into the guard assembly and may not actually project from the side over the treads. That configuration would provide the designer flexibility in the stairway width, including the treads. Although the code does not require a handrail on each side of a residential stairway, if two handrails are installed, one on each side, then the width of the stair treads and landings can be reduced to as little as 27 inches (698 mm) (see Example 6-7).

Generally speaking, it is better to provide the full 36-inch (914 mm) width for the entire height of the stairway, as handrail design does not always work out as planned when it comes to incorporating them into the guard assembly (details of this are provided in Chapter 7). Similarly, landings must still be 36 inches (914 mm) in the direction of travel, so a stairway that turns 90 degrees (1.57 rad) at a landing would be somewhat strange with a reduced width. Utilizing a reduced stairway width should only be done with prior approval of the local code official when there are practical limitations, such as when remodeling an existing structure. An exception in the code to the minimum headroom at stairways is provided for spiral stairways where it allows a slight reduction for required headroom (see the discussion of Section R311.7.9, later in this chapter).

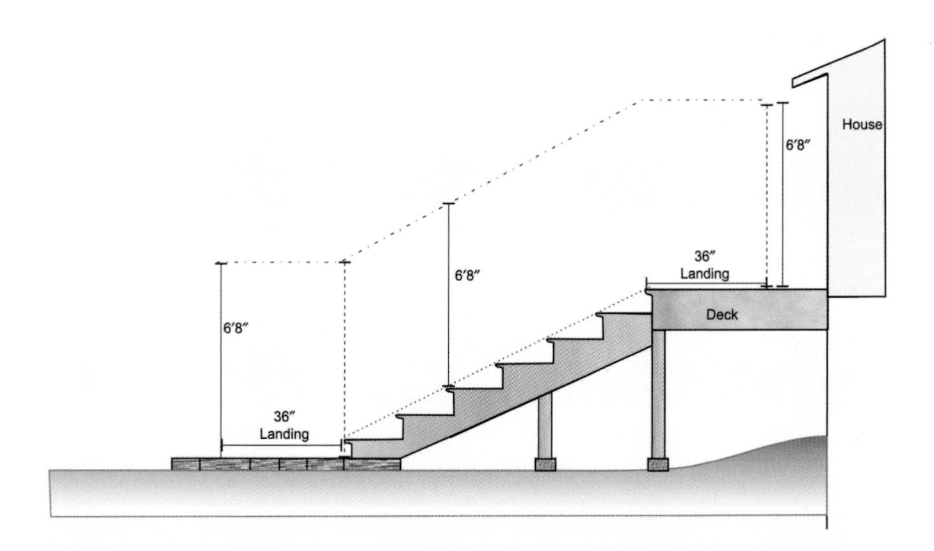

Example 6-6: When a stairway accesses a deck, patio or other large walking surface, a portion of the large walking surface either at top and/or bottom is considered the landing for the stairs. While not identified or distinct, the area of the large walking surface encompassing the width of the stair and 36 inches (914 mm) in depth from the top or bottom nosing is part of the stairway and must be provided with the required headroom.

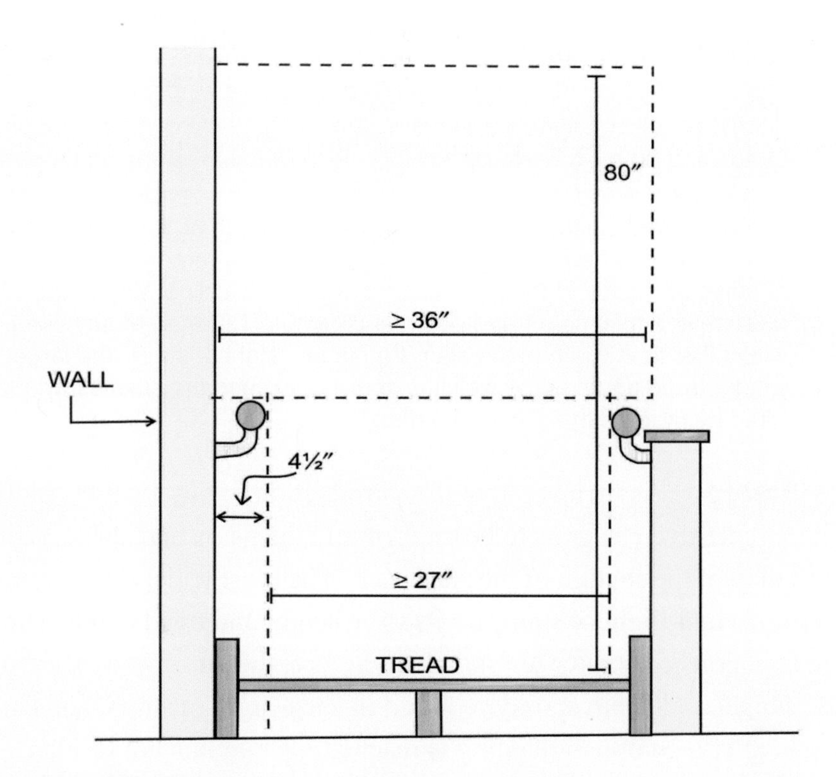

Example 6-7: Generally, stairways must be at least 36 inches (914 mm) wide. When handrails are provided the width at and under the handrails may be reduced by $4^1/_2$ inches (114 mm) on each side. While not necessarily the intention, this reduced width may allow for flexibility in design by using narrower treads when the handrails are to the side of the stair treads and not over them.

Part Two: Stair Geometry

R311.7.3 Walkline. The walkline across winder treads shall be concentric to the curved direction of travel through the turn and located 12 inches (305 mm) from the side where the winders are narrower. The 12-inch (305 mm) dimension shall be measured from the widest point of the clear stair width at the walking surface of the winder. If winders are adjacent within the flight, the point of the widest clear stair width of the adjacent winders shall be used.

R311.7.4 Stair treads and risers. Stair treads and risers shall meet the requirements of this section. For the purposes of this section all dimensions and dimensioned surfaces shall be exclusive of carpets, rugs or runners.

R311.7.4.1 Riser height. The maximum riser height shall be $7^3/_4$ inches (196 mm). The riser shall be measured vertically between leading edges of the adjacent treads. The greatest riser height within any flight of stairs shall not exceed the smallest by more than $^3/_8$ inch (9.5 mm).

R311.7.4.2 Tread depth. The minimum tread depth shall be 10 inches (254 mm). The tread depth shall be measured horizontally between the vertical planes of the foremost projection of adjacent treads and at a right angle to the tread's leading edge. The greatest tread depth within any flight of stairs shall not exceed the smallest by more than $^3/_8$ inch (9.5 mm). Consistently shaped winders at the walkline shall be allowed within the same flight of stairs as rectangular treads and do not have to be within $^3/_8$ inch (9.5 mm) of the rectangular tread depth.

Winder treads shall have a minimum tread depth of 10 inches (254 mm) measured between the vertical planes of the foremost projection of adjacent treads at the intersections with the walkline. Winder treads shall have a minimum tread depth of 6 inches (152 mm) at any point within the clear width of the stair. Within any flight of stairs, the largest winder tread depth at the walkline shall not exceed the smallest winder tread by more than $^3/_8$ inch (9.5 mm).

Discussion: As mentioned in the introduction to this chapter, stairs require repetitive elements to be uniform in dimension and shape, so as not to disrupt the rhythm of the stair user. The most widely recognized repetitions in a flight of stairs are its riser height and tread depth. The above sections only refer to "stairs"; therefore, the uniformity of the tread height and depth is only required in each flight of stairs, and can change after a compliant landing is reached. While it is not ideal, a stairway that consists of two flights of stairs, divided by a landing, can have dramatically different riser height and tread depth between the two (see Example 6-8). It is assumed that the rhythm created by the stair user has been interrupted by the landing, and a new rhythm will begin at the next stair.

In each individual set of stairs, the tallest riser height and the shortest riser height cannot differ by more than $^3/_8$ inch (9.5 mm). Likewise, the longest tread depth cannot differ from the shortest by more than $^3/_8$ inch (9.5 mm).

Unlike commercial stairs, there is no minimum riser height for an IRC step; there is only a maximum. A riser cannot exceed $7^3/_4$ inches (196 mm), but could be as little as 2 inches (51 mm) or less. Shallow risers are a design flexibility afforded by the IRC, but anything less than the commercial minimum of 4 inches (102 mm), as provided in the IBC, may become a trip hazard if not easily distinguishable as a step (see Example 6-9). Inversely, the depth of a tread has no maximum required dimension, only a minimum. The tread depth is measured horizontally and at a right angle from the nosing of the tread in question to the nosing of the tread or landing above it. The minimum tread length, or depth, is 10 inches (254 mm); however, that is dependent on the existence of a compliant tread nosing projection (see next discussion). A maximum tread depth does not exist, yet once 36 inches (914 mm) is reached, the tread would be considered a landing and the stair uniformity will be completed. With that said, a stair with deep treads less than 36 inches (914 mm), such as a four-deck board step approximately 24 inches (610 mm) deep, would still need to be uniform to the $^3/_8$-inch maximum (9.5 mm).

Winder treads (see definitions) are a bit different, as there are two tread depth measurements that must be considered. At the narrowest edge of a winder tread, the side of the tread the stair turns toward, a minimum 6-inch (152 mm) depth must be provided. This depth is just a minimum, and does not have to have uniformity in relation to the other treads. The tread depth of a winder tread is measured at a point 12 inches (304 mm) from the narrower edge of the tread within the clear width of the stairs and at the level of the treads (see Example 6-10). This point is referred to as the "walkline," as it is in the turn made by the winder treads and the most likely place people will choose to walk. While this provision appears to make perfect sense, it may become difficult to apply consistently in construction where various guard and handrail designs affect where a person can actually walk. When planning the geometry of wider treads, careful attention must be given to how the guard or handrail system will be installed. The 6-inch (152 mm) minimum tread depth at the tread edge and the location of the walkline must be measured with consideration to the location of the guard. Whether a guard assembly is built off the stairs, on the side, or over the stairs at the side, will affect how these two requirements are measured. The maximum distance that handrails can project over the treads is discussed in greater detail in Chapter 7.

There is one other nuance in regard to the uniformity of stairs that include winder treads. A single stair may include both parallel treads and winder treads. When this occurs, the tread depth of the winder treads, measured as described above, must maintain a $^{3}/_{8}$-inch (9.5 mm) uniformity between the winders, but not with the parallel treads. Likewise, the parallel treads' depths are only required to be uniform with the other parallel treads, and not the winders. In many cases two or more winder treads may be used to create a turn in a single stair. These winder treads often have tread depths greater than the parallel treads' depths—all within a single stair. The rational for allowing this lack of uniformity is similar to the existence of an intermediate landing between two stairs; the rhythm is interrupted by the turn already, thus the uniformity is not as necessary (see Example 6-11).

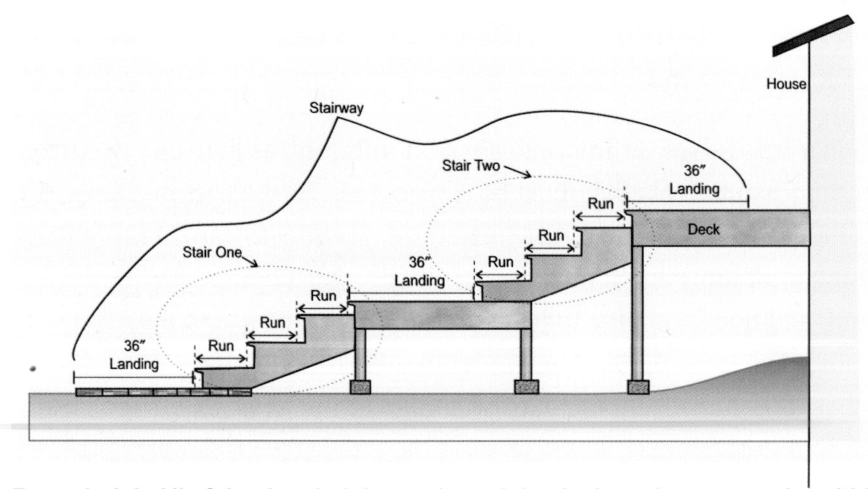

Example 6-8: All of the riser heights and tread depths in stair one must be within $^{3}/_{8}$ inch (9.5 mm) of one another, respectively, yet can be completely different than the riser height and tread depth in stair two. While not necessarily comfortable to walk, this single stairway can be constructed of dramatically different individual stairs, regardless of whether the landing in the middle is part of a bigger deck or not.

Example 6-9: The step between this concrete patio and deck is only about 2 inches (51 mm) high. While this may become a trip hazard, as it is difficult to distinguish, there is no minimum riser height required in the IRC. However, this small single step is considered a stair and a stairway and is subject to all related provisions, such as safety glazing and illumination.

Example 6-10: Winder treads break up the "square" look of traditional stairs but can be challenging to lay out and construct in accordance with all the geometric requirements of the IRC.

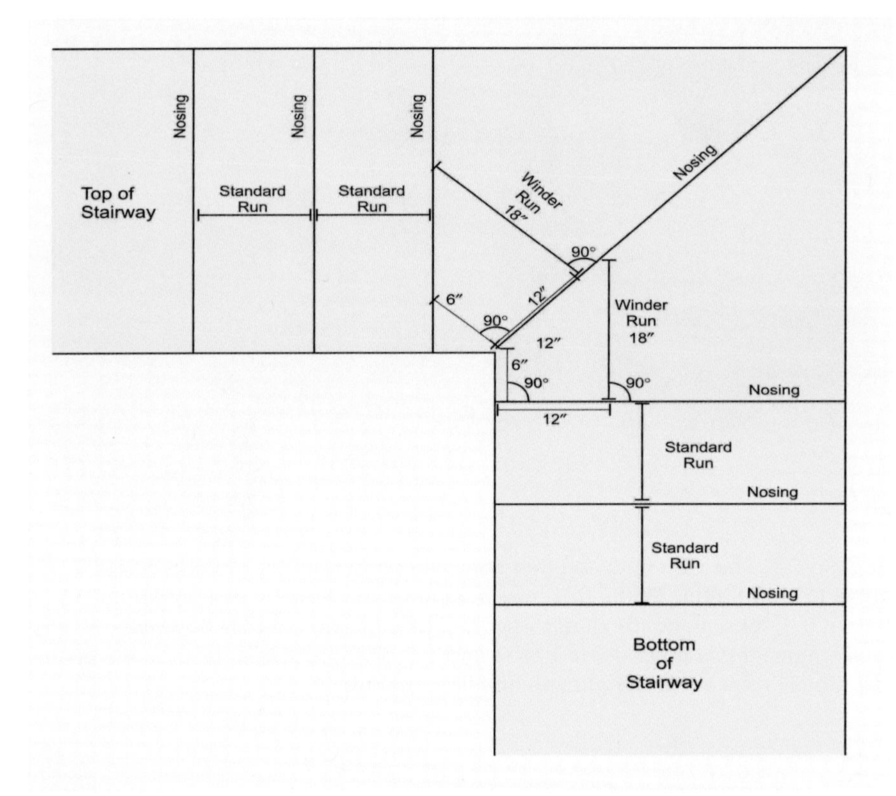

Example 6-11: Combining winder treads with parallel treads to make a turn in a single stair is not an easy task. Laying out the location of the winder nosings with respect to the minimum 6 inch (152 mm) tread depth and the tread uniformity at the 12-inch (305 mm) walkline requires preparation and planning. This example details the geometry of creating a 90 degree (1.57 rad) turn using two winder treads.

Side Note: Remember to cut off the thickness of the decking material from the bottom of the stair stringers when supported by the finished landing surface. A concrete landing, for example, does not receive tread material, as the steps above do, and thus the bottom riser of stairs is often constructed in error, and results in a taller rise than the others by the thickness of the tread material.

R311.7.4.3 Profile. The radius of curvature at the nosing shall be no greater than $^{9}/_{16}$ inch (14 mm). A nosing not less than $^{3}/_{4}$ inch (19 mm) but not more than $1^{1}/_{4}$ inches (32 mm) shall be provided on stairways with solid risers. The greatest nosing projection shall not exceed the smallest nosing projection by more than $^{3}/_{8}$ inch (9.5 mm) between two stories, including the nosing at the level of floors and landings. Beveling of nosings shall not exceed $^1/_2$ inch (12.7 mm). Risers shall be vertical or sloped under the tread above from the underside of the nosing above at an angle not more than 30 degrees (0.51 rad) from the vertical. Open risers are permitted, provided that the opening between treads does not permit the passage of a 4-inch diameter (102 mm) sphere.

Exceptions:

1. A nosing is not required where the tread depth is a minimum of 11 inches (279 mm).
2. The opening between adjacent treads is not limited on stairs with a total rise of 30 inches (762 mm) or less.

Discussion: The required geometry of stairs is explained further in this section. The leading edge of a stair tread is referred to as the nosing (see definitions). However, at the top of a stair the landing or floor surface must also provide a nosing for the stair. The nosing itself cannot have a radius curve of more than $^9/_{16}$ inch (14 mm), or a bevel of more than $^1/_2$ inch (12.7 mm), measured along the angle of the bevel. Often, when ascending stairs, people place the balls of their feet on or near the nosing, and a significant bevel or radius can cause the foot to slip off the tread (see Example 6-12).

Essentially, the IRC wants every tread to provide at least $10^3/_4$ inches (273 mm) of depth for placement of the foot. This is created by the minimum 10-inch (254 mm) tread depth, described in the previous section, and the requirement for a minimum $^3/_4$ inch (19 mm) nosing projection. This projection, which cannot exceed $1^1/_4$ inches (32 mm), allows a foot to extend under the tread above, providing a bit more depth for foot placement than the tread depth provides. Unlike riser height and tread depth uniformity, the uniformity of the stair nosings cannot vary after a landing. All nosings in a single stairway containing one or more flights must maintain $^3/_8$-inch (9.5 mm) uniformity in their projection distances.

The projection can be eliminated, however, if depth for foot placement is provided in another manner, such as completely open risers on stairs less than 30 inches (762 mm) above grade, or treads with at least an 11 inch (279 mm) tread depth. When the overall height of a flight of stairs exceeds 30 inches (762 mm), regardless of the number of individual risers, the risers are required to prevent the passage of a 4-inch (102 mm) diameter sphere. This is consistent with the allowable opening of guard systems as discussed in Chapter 7. Most often this blockage is created by a solid riser. Similar to guards, the opening limitation of risers on stairs is intended to provide fall protection for children. The total height of the stair is intended to be measured from the grade below, not an intermediate landing. An upper stair in a stairway divided by landings may individually have a total height of less than 30 inches (762 mm), as measured from the intermediate landing, yet the distance to the ground below from between the risers could be considerably higher (see Example 6-13).

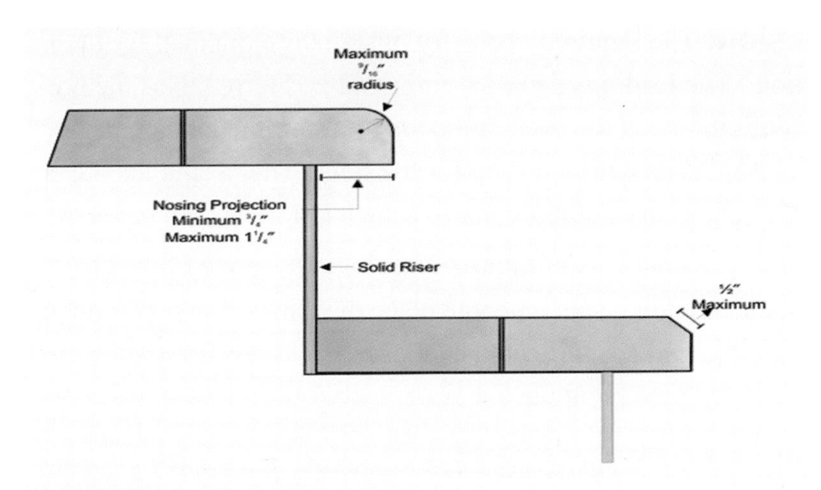

Example 6-12: Rounding or beveling the nosing of a tread can add a nice touch to a custom deck but can inadvertently create a situation where the tread does not comply with the code.

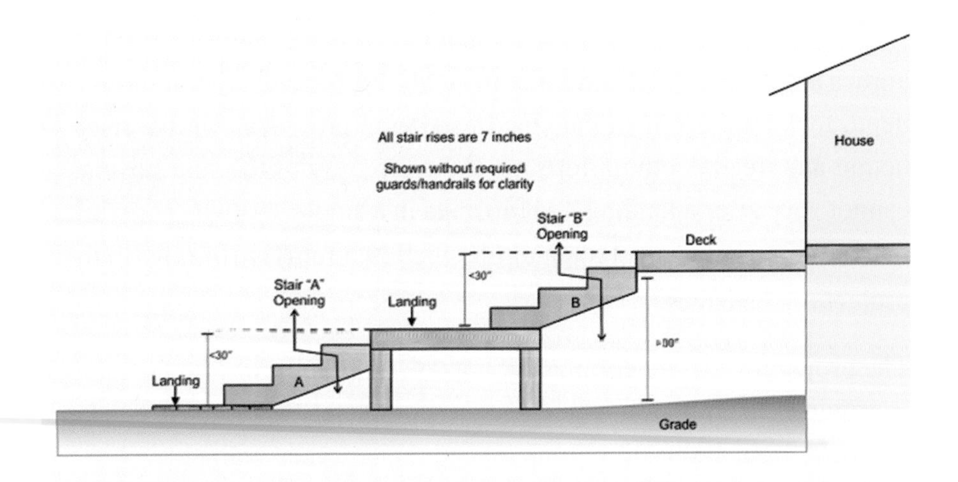

Example 6-13: In this stairway, each stair has a total rise of only 28 inches (711 mm) and would seemingly apply to the second exception to riser opening restrictions. In accordance with the intent of Section R311.7.4.3, Stair B, although less than a 30-inch (762 mm) total rise, would result in a distance of greater than 30 inches (762 mm) to the grade below. In this example, the exception would apply only to Stair A.

Side Note: In regions that receive snowfall, solid risers can create a dangerous condition if snow is not cleared and is packed into the inside corners of the risers and treads. For stairs with a total height of 30 inches (762 mm) or less the IRC does not require risers that prohibit the passing of a 4-inch (102 mm) diameter sphere. For stairs greater than 30 inches (762 mm) above grade, a more slender riser board, held 3 $^{3}/_{4}$ inches (95 mm) above the tread below, can be used to both reduce the opening size while allowing room for snow to push through.

R311.7.5 Landings for stairways. There shall be a floor or landing at the top and bottom of each stairway.

Exception: A floor or landing is not required at the top of an interior flight of stairs, including stairs in an enclosed garage, provided a door does not swing over the stairs. A flight of stairs shall not have a vertical rise larger than 12 feet (3658 mm) between floor levels or landings. The width of each landing shall not be less than the width of the stairway served. Every landing shall have a minimum dimension of 36 inches (914 mm) measured in the direction of travel.

R311.7.6 Stairway walking surface. The walking surface of treads and landings of stairways shall be sloped no steeper than one unit vertical in 48 inches horizontal (2-percent slope).

Discussion: An intermediate landing between two flights of stairs can serve as the top landing for one and the bottom landing for the other. Each individual flight cannot exceed 12 feet (3658 mm) in total height between the required landings (see Example 6-14).

Landings must be at least equal to the width of the stairs, regardless of the magnitude of that width, and at least 36 inches (914 mm) in the direction of travel, measured from the nosing of the last tread (see Examples 6-15, 6-16 and 6-17). There is no material specified by the IRC for stair landings; however, the intent is for a solid, stable material with an identifiable and measurable slope. Properly installed concrete, pavers, flagstone, brick or decking may all be considered suitable as landings, provided they comply with the size and slope requirements (see Example 6-18). Landings and stair treads cannot have an individual slope greater than $^{1}/_{4}$ inch in 12 inches (2-percent slope) in any direction, either parallel or perpendicular to the nosing.

Example 6-14: Although serving a deck at a restaurant and thus regulated under the IBC, the same provision applies for decks constructed under the IRC. The total rise of this stairway is greater than 12 vertical feet (3658 mm); thus an intermediate landing was required to create two individual "stairs" or "flights" within the stairway, each less than the maximum 12-foot (3658 mm) total rise.

Example 6-15: Regardless of the shape of a stair, a landing is required for its entire width.

Example 6-16: The landing at the bottom of this stairway is compliant and required, and while it could be argued as excessive, it is merely the result of an excessively designed stairway.

Example 6-17: Unfortunately this landing was poured without respect to the nosing of the last tread. While it was originally poured at 36 inches (914 mm), the placement of the stairs on the landing effectively shortened the minimum required dimension and a correction was required. The minimum landing distance in the direction of travel must be measured from the nosing of the last tread.

Example 6-18: The landings pictured above satisfy the size and slope requirements of the IRC for landings, and are acceptable.

R311.7.8 Illumination. All stairs shall be provided with illumination in accordance with Section R303.6.

R303.6 Stairway illumination. All interior and exterior stairways shall be provided with a means to illuminate the stairs, including the landings and treads. Interior stairways shall be provided with an artificial light source located in the immediate vicinity of each landing of the stairway. For interior stairs the artificial light sources shall be capable of illuminating treads and landings to levels not less than 1 foot-candle (11 lux) measured at the center of treads and landings. Exterior stairways shall be provided with an artificial light source located in the immediate vicinity of the top landing of the stairway. Exterior stairways providing access to a *basement* from the outside *grade* level shall be provided with an artificial light source located in the immediate vicinity of the bottom landing of the stairway.

> **Exception:** An artificial light source is not required at the top and bottom landing, provided an artificial light source is located directly over each stairway section.

R303.6.1 Light activation. Where lighting outlets are installed in interior stairways, there shall be a wall switch at each floor level to control the lighting outlet where the stairway has six or more risers. The illumination of exterior stairways shall be controlled from inside the *dwelling* unit.

> **Exception:** Lights that are continuously illuminated or automatically controlled.

Discussion: Section R303.6 provides two requirements for illumination of exterior stairways, and each one is separately applicable. The leading statement of this section requires that all stairs, landings and treads of all exterior stairways be provided with a "means to illuminate." This "means" can come from any source or location, be it a flood light mounted to the existing home, or small low-voltage lights dispersed throughout all the stairway components (see Examples 6-19 and 6-20). Stairs are inherently dangerous, and therefore the means to illuminate them completely may help inhibit an after-dark slip and injury.

Further in the section there is a specific requirement for the location of an "artificial light source" in the immediate vicinity of the top landing. The specific requirement for a light source only at the top landing of the stairway does not eliminate the requirement for a "means to illuminate" all the stairway components. Depending on the layout of the stairs, and the location, size, type and wattage of the required upper landing light, the stairway may be illuminated in its entirety by a single light, or it may require additional light sources. When low-voltage lights are installed throughout a stairway as the "means to illuminate," at least one light fixture must be placed at the top landing.

The illumination of exterior stairways, regardless of the quantity or location of the fixtures, must be controllable from inside the dwelling. This does not mean that other illumination cannot be provided separate from the stairway illumination and controlled elsewhere, nor does it prohibit three-way or four-way switches to allow control of the stairway illumination from additional locations.

The exception to this requirement is convenient for low-voltage lighting systems commonly seen in deck construction. Stairway lighting not switched inside must be continuously illuminated or automatic. For continuous illumination, a timer or photo cell can be used that would turn the lights on at dusk and off at dawn. Many low-voltage transformers allow multiple settings of the timer and photo cell in combination and can be set to turn the lights off after a predetermined period of time. While this saves the electricity of running lights throughout the entire night, it does not satisfy the IRC requirements. Motion sensors can also be used that would turn the lights on automatically when sensing a stair user's approach to either a top or bottom landing. The intent of this section is to allow someone to illuminate the stairs at any hour of the night, either from inside the dwelling or automatically.

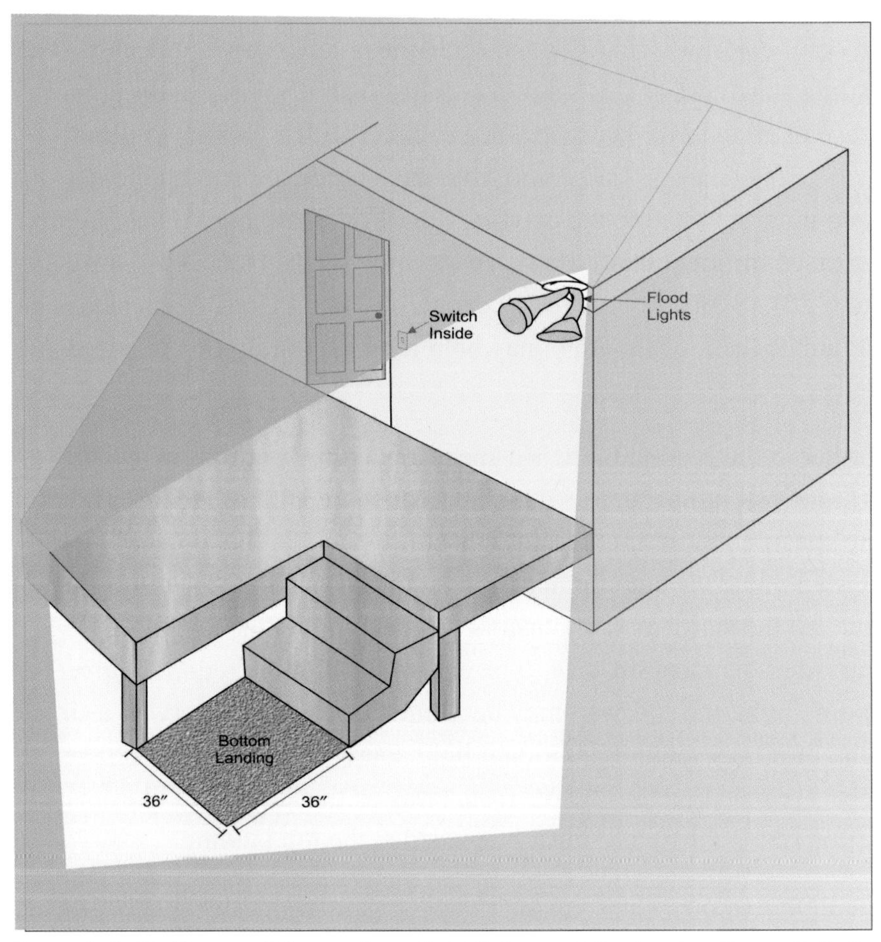

Example C-10: The small low voltage light fixtures at the top landing satisfy the requirement for the location of a light source, yet are not sufficient to provide a "means to illuminate" the entirety of the stairway. The remaining recessed step lights provide the additional illumination required.

Example 6-20: If a single light located in the "immediate vicinity of the top landing" is capable of illuminating the entire stairway, no other "means to illuminate" is necessary.

R311.7.9 Special stairways. Spiral stairways and bulkhead enclosure stairways shall comply with all requirements of Section R311.7 except as specified below.

R311.7.9.1 Spiral stairways. Spiral stairways are permitted, provided the minimum clear width at and below the handrail shall be 26 inches (660 mm) with each tread having a $7^1/_2$-inch (190 mm) minimum tread depth at 12 inches (914 mm) from the narrower edge. All treads shall be identical, and the rise shall be no more than $9^1/_2$ inches (241 mm). A minimum headroom of 6 feet 6 inches (1982 mm) shall be provided.

Discussion: The IRC recognizes spiral stairways as a historically common construction feature throughout the US, and allows their continued construction and use in residential applications. Spiral stairways are sometimes designed into deck construction, although they are usually premanufactured to fit the specific site conditions and delivered as a complete unit with minimal on-site assembly. Regardless of where the stairs are constructed, if they are part of a deck installation, they will be required to comply with the IRC provisions during the evaluation of the deck. Clearly understanding that spiral stairs are not intended for moving large objects, but rather just people, the IRC allows reduced width and headroom clearances, as well as an increased riser height and decreased tread depth limitation.

Unlike winder treads, spiral stairways do not require the minimum 6-inch (152 mm) tread depth on the narrower side of the tread. As described in the definition provided at the start of this chapter, the treads must be of identical shape and must rotate circularly around a center post (see Example 6-21).

Example 6-21: As shown in this photo, a spiral stairway is a very specific type of stairway, as explained in the IRC definition.

Part Three: Ramps

R311.8.1 Maximum slope. Ramps shall have a maximum slope of 1 unit vertical in 12 units horizontal (8.3 percent slope).

> **Exception:** Where it is technically infeasible to comply because of site constraints, ramps may have a maximum slope of one unit vertical in eight horizontal (12.5 percent slope).

R311.8.2 Landings required. A minimum 3-foot-by-3-foot (914 mm by 914 mm) landing shall be provided:

1. At the top and bottom of ramps.
2. Where doors open onto ramps.
3. Where ramps change direction.

Discussion: From a structural perspective, ramps are nothing more than out-of-level decks, and do not require any special consideration in regard to their framing or live load requirements. The information in Chapter 4 can be utilized for construction of these long, narrow, sloping "decks." The ramp provisions in the IRC are intended to satisfy a general desire for a ramp at a private residence, but not at commercial locations. The IBC and ANSI A117.1 provide much more detailed requirements for ramp construction in public settings, intended for maximum safety of disabled individuals.

Where ramps have to be considered differently from decks is in the sloping part of their description. The leading statement in this section is for ramps to not exceed a slope of one unit vertical in 12 units horizontal (see Example 6-22). To fully grasp the results of a ramp with that slope, consider the following: To make a 24-inch (610 mm) vertical change in elevation with a ramp, the ramp would have to be 24 feet (7315 mm) long. The math of this provision allows for easy memorization; for each inch of vertical rise, you need a foot of horizontal distance.

There is only one way to increase the slope of a ramp, and that would be through the approval of the building official, based on the claim that the site conditions do not allow for the ramp length in order to make the maximum required slope. That approval needs to be sought prior to construction and plan submittal, not afterwards, as the general design of the deck will likely be considered when evaluating the "site conditions." If a very large deck minimizes the space for ramp, it will likely not be considered a site restraint, as the deck could be designed differently or smaller. Similarly, a desire for less sacrifice of yard area will likely not be considered a site restraint; if there is space, the ramp must slope at least 1 unit in 12. However, if you were to gain the approval for an increased slope, the same 24-inch (610 mm) rise would be able to be shortened from 24 feet (7315 mm) to 16 feet (4877 mm). While this is quite a reduction, 16 feet (4877 mm) of ramp definitely needs to be planned into the original design.

Much like stairways, landings are required at the top and bottom of all ramps to provide a safe transition from the sloping surface to the level surface. The IRC only requires landings to be a minimum of 3 feet by 3 feet (914 mm by 914 mm) square. However, if a ramp is wider than 3 feet (914 mm), the intent of the code is to provide a landing equal to the ramp's width. Wherever a ramp makes a change of direction, a landing is required. Landings are also required where doors open onto a ramp, yet this provision is more specific to doors and is covered more specifically elsewhere in the IRC (see Chapter 2). Just as for stairs, landings for ramps and cross-slopes of ramps cannot have a slope exceeding $^1/_4$ inch in 12 inches. Handrail requirements for ramps are a result of the slope of the ramp, and are discussed in Chapter 7.

Example 6-22: This ramp has a slope exceeding 1 in 12 and would not qualify for "site restraints" as there is a large unused yard in the photo; it is noncompliant. However, the ramp is provided with a compliant landing at the top and at the corner of the deck where the ramp changes direction.

Chapter 7: Guards and Handrails

Introduction

Staying upright and on our feet is a simple way to limit accidents, and the IRC has provided minimum construction standards expected to do just that. Guards and handrails are the components intended to provide this security to our stability, yet they do so in different ways and are different construction elements. A guard may contain an acceptable handrail within its makeup, depending on its design, but a handrail alone can never be a guard; they are not one and the same.

Guards, often seen in the form of rails, are intended to block the edge of a deck, stairway or other raised walking surface from an inadvertent fall. For a child, adult or even the family dog, guards will inhibit a fall over the edge, but they can also provide other functions to a deck. As stated in the IRC definition for "guard," they can be any "building component or system of building components." Construction elements of any shape, size, material or function, which can satisfy the minimum geometric and structural requirements provided in this chapter, can function as a compliant guard when properly installed. If properly designed, a wall, planter box, bench-seat back, built-in barbeque or the traditional rails and balusters assembly can all act as guards.

Handrails, while still providing fall protection, are intended to work in a different manner. Rather than obstructing an accidental fall from the edge of a walking surface, a handrail is meant for your purposeful use while traversing irregular walking surfaces, such as ramps and stairways. To perform this function, handrails have more specific requirements in height and cross-section than guards. A handrail is a horizontally sloping rail placed down the length of a ramp or stairway to allow an individual to grasp for assistance, to either inhibit a fall from occurring or to brace oneself when it does occur.

Definitions

GUARD. (IRC) A building component or a system of building components located near the open sides of elevated walking surfaces that minimizes the possibility of a fall from the walking surface to a lower level.

HANDRAIL. (IRC) A horizontal or sloping rail intended for grasping by the hand for guidance or support.

Newel Post. (McGraw-Hill) A post at the end of a straight stairway or on a landing.

Part One: Handrails

R311.7.7 Handrails. Handrails shall be provided on at least one side of each continuous run of treads or flight with four or more risers.

R311.8.3 Handrails required. Handrails shall be provided on at least one side of all ramps exceeding a slope of one unit vertical in 12 units horizontal (8.33-percent slope).

Discussion: As discussed in the introduction, handrails are an incredibly simple and inexpensive method of providing additional safety to people traversing stairways and ramps. Handrails have become so ingrained in our built environment that for many people the extension of their hand while approaching stairs has become a reflex. However, as the decking industry continues to turn out unique and creative outdoor living spaces, cascading stairways, winding and circular stairways, and monumental stairways are often designed without considering how the required handrail will fit. Likewise, many custom and manufactured guard systems do not have compatible handrails included in their system, leaving a last-minute, unplanned handrail, often of a different material, as an unattractive finish to the project.

Any and every time you design a stairway, you must keep in the forefront of your creativity a single handrail on one side of each stair with four or more risers; two handrails are never required by the IRC. While this section refers to "a continuous run of treads" it is essentially referring to the definition of "flight" or "stair" (see Chapter 6). If a stairway was sectioned as individual stairs with no more than three risers each and separated by compliant landings, an infinitely high stairway could be designed without the use of a handrail. Presumably, a stairway user can rest at each landing between the short flights of stairs, and in an accidental fall, the tumble would presumably stop at the next landing below (see Examples 7-1 and 7-2). In ramp construction, a handrail is only required on one side when the slope is greater than one unit vertical for every 12 units horizontal, no matter the overall length or rise of the ramp itself. However, a slope that steep can only be approved when a lower-slope ramp is technically infeasible due to site constraints (see Chapter 6).

Example 7-1: This stairway consists of three separate stairs divided by compliant landings. Each stair has only three risers; therefore, no handrail is required. With intermediate landings provided, an accidental fall will likely end at the nearest landing, rather than down the entire stairway. A stair user may also stop and rest at each landing, minimizing the need to grasp a handrail while using the short stair sections.

Example 7-2: A similar entry stairway as the previous example, yet this one is designed in a manner that requires a handrail. While the run of the individual stair treads are unusually deep, they are not compliant landings. This stairway is made up of a single stair, with more than three risers, and thus a single handrail is required.

R311.7.7.1 Height. Handrail height, measured vertically from the sloped plane adjoining the tread nosing, or finish surface of ramp slope, shall be not less than 34 inches (864 mm) and not more than 38 inches (965 mm).

R311.8.3.1 Height. Handrail height, measured above the finished surface of the ramp slope, shall be not less than 34 inches (864 mm) and not more than 38 inches (965 mm).

Discussion: The code allows handrails to be installed within a small range of heights. This provides beneficial flexibility in the alignment and connection of the handrail to a level guard, as well as allowing some room for variation in the installation. The height is measured vertically from the surface of a ramp or from the nosing of the stair treads (see Example 7-3). It is important to realize that measuring the height from any part of the stairway other than the nosing may result in a non-compliant installation. Likewise, if the handrail is not parallel to the angle of the stair stringer, the height may be correct at one end of the stair but not at the other end of the stair. The code does not, however, specifically require the handrail to be parallel to the stair stringer or ramp surface, as long as all portions are within the prescribed range of heights.

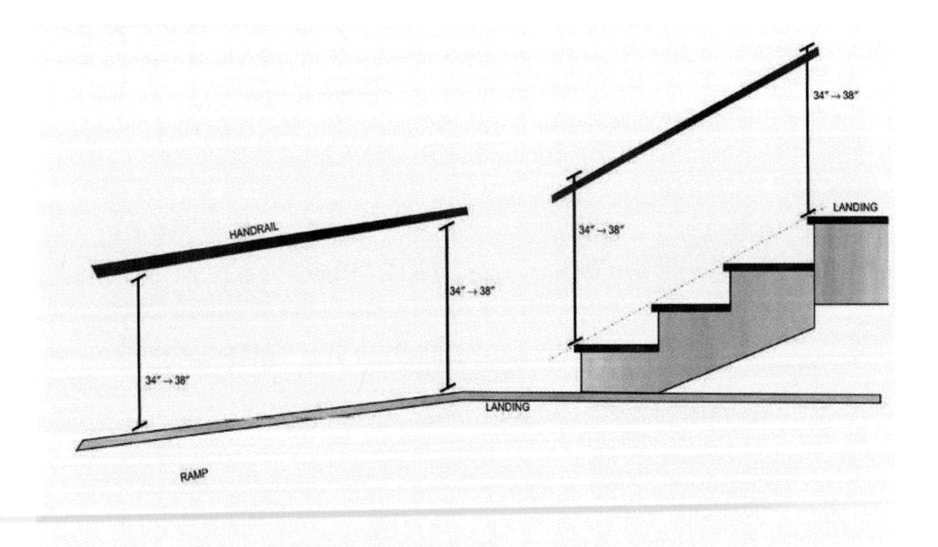

Example 7-3: Unlike guards, handrails have a minimum and maximum height requirement. No matter the style and shape of the graspable rail, the height is always measured to the top surface of the rail.

R311.7.7.2 Continuity. Handrails for stairways shall be continuous for the full length of the flight, from a point directly above the top riser of the flight to a point directly above the lowest riser of the flight. Handrail ends shall be returned or shall terminate in newel posts or safety terminals. Handrails adjacent to a wall shall have a space of not less than $1^1/_2$ inch (38 mm) between the wall and the handrails.

Exceptions:

1. Handrails shall be permitted to be interrupted by a newel post at the turn.
2. The use of a volute, turnout, starting easing or starting newel shall be allowed over the lowest tread.

Discussion: Once a person begins traversing a stair, referred to in this section as a "flight," and they grasp the handrail, they must be provided the handrail for the duration of their trip to the next landing. From the top of the stair, the handrail must begin at a point vertically above the first nosing, the one that is at the same height as the floor/landing. It must continue, uninterrupted by posts or other obstructions, until it reaches a point vertically above the last nosing of the stair, at the level of the bottom tread (see Example 7-4). Although commonly extended beyond the last riser, and required to do so in commercial structures, the IRC does not require an extension of the handrail beyond the top or bottom risers. Throughout this handrail length, a stair user must be able to slide their hand up or down the rail without the need to let go for any obstruction. However, when a single flight of steps contains winder treads that create a turn in the stair (see Chapter 6), an exception allows the handrail continuity to be interrupted by a newel post (see Example 7-5).

Both ends of a handrail must be designed so that they return to the wall or terminate in newel posts or safety terminals. Model building codes have historically provided this safety provision to fire personnel to inhibit a handrail from catching the equipment on their backs or the hoses they are pulling up a stairwell. For the everyday deck user, the required height of handrails places them in an ideal location to snag your pants pockets, purse strings or the bottom of your shirt if the handrail end is left sticking out (see Example 7-6). When a handrail is mitered and joined so that it turns and terminates into a post at the side of the stairs, it minimizes the opportunity for clothes to be caught (see Examples 7-7 and 7-8). Depending on the location of the posts, the handrail may not need to make the return to the post; it may continue straight and terminate directly into the post. While not so common with typical deck handrail materials, a safety terminal may also be used. As commonly seen in interior handrails, the use of a volute, turn-out or other ornamental termination of the handrail, which creates a large enough roll, spiral or downward curve is an acceptable method of inhibiting the handrail's ability to snag clothes or other items. For deck construction, steel or wrought iron handrails can easily be designed to include safety terminals; this is not as common or easy using a handrail with mitered joints (see Example 7-9).

The second exception in this section allows the safety termination or termination into a post to occur over the last tread, allowing the continuity of the handrail to be slightly short of the nosing of the last tread.

When a handrail is attached to the side of a wall, post or other mounting surface, there must be a clear distance of at least $1^1/_2$ inches (38 mm) between the rail and the wall to allow for fingers to fully grasp the rail.

Example 7-4: Had the top rail of this stair guard been complaint as a graspable rail (see R311.7.7.3, next discussion) it still would not comply with the required continuity, due to the center post. A continuous and properly terminated handrail had to be added to one side of the guard assembly.

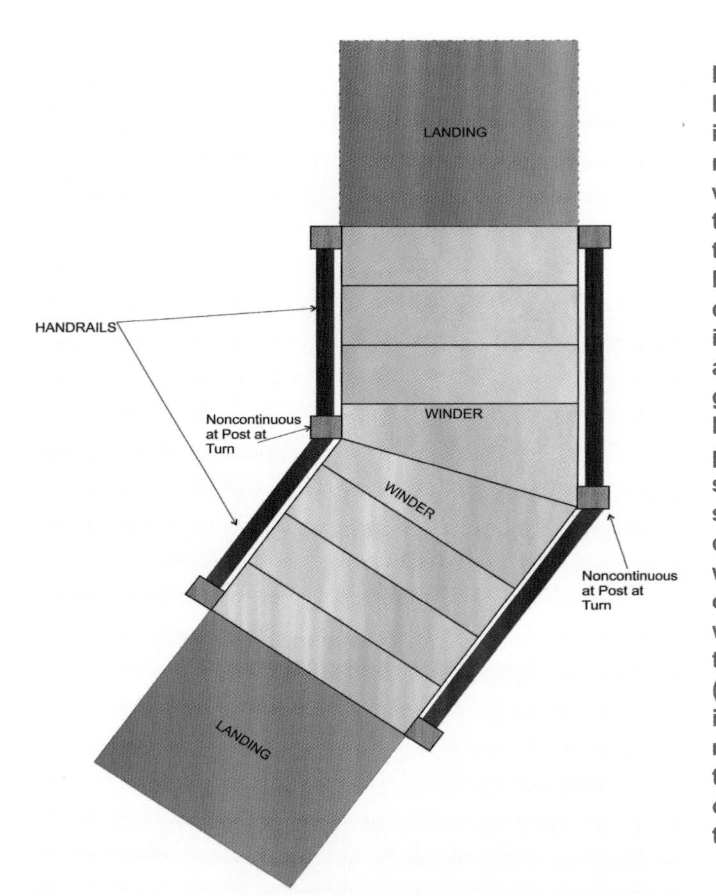

Example 7-5: A handrail can be interrupted by a newel post where winder treads cause a turn in the stair. For increased design flexibility, the IRC now allows the single required handrail to be placed on either side of winder stairs. However, considering the walkline method of designing winder stair tread geometry (see Chapter 6), it would be more functional to place the rail on the inside of the stair turn.

Example 7-6: This steel handrail was installed without a return to the post or safety terminal. In an attempt to correct the hazard in accordance with the "intent and purpose" of the IRC, composite decking "deflectors" were installed at the top and bottom in order to eliminate the ability for the rail to snag people's clothing or accessories. The approval of this "fix" alternative would be the authority of the building official.

Example 7-7

Example 7-8: Handrail terminations can vary depending on the material used. A curved metal handrail or mitered composite handrail, whether returning to a post or a wall, will both serve the intent of this IRC requirement.

Example 7-9: These powder-coated steel guards, functioning as handrails, are provided with a termination design that serves as a compliant safety terminal.

R311.7.7.3 Grip-size. All required handrails shall be of one of the following types or provide equivalent graspability.

1. Type I. Handrails with a circular cross section shall have an outside diameter of at least $1^1/_4$ inches (32 mm) and not greater than 2 inches (51 mm). If the handrail is not circular it shall have a perimeter dimension of at least 4 inches (102 mm) and not greater than $6^1/_4$ inches (160 mm) with a maximum cross section of dimension of $2^1/_4$ inches (57 mm). Edges shall have a minimum radius of 0.01 inch (0.25 mm).
2. Type II. Handrails with a perimeter greater than $6^1/_4$ inches (160 mm) shall have a graspable finger recess area on both sides of the profile. The finger recess shall begin within a distance of $^3/_4$ inch (19 mm) measured vertically from the tallest portion of the profile and achieve a depth of at least $^5/_{16}$ inch (8 mm) within $^7/_8$ inch (22 mm) below the widest portion of the profile. This required depth shall continue for at least $^3/_8$ inch (10 mm) to a level that is not less than $1^3/_4$ inches (45 mm) below the tallest portion of the profile. The minimum width of the handrail above the recess shall be $1^1/_4$ inches (32 mm) to a maximum of $2^3/_4$ inches (70 mm). Edges shall have a minimum radius of 0.01 inch (0.25 mm).

Discussion: The cross-sectional shape of a handrail must be designed so that hands of all sizes are capable of achieving a secure grasp. The charging statement of this section allows the user to choose from two different specific cross-sectional designs. However, it is important to note that this section reinforces the IRC flexibility in using alternative methods that will provide an equivalent function. This section clearly reminds the code user that the intent and purpose is to provide a handrail that can be grasped, regardless of whether the specific and detailed measurements and geometry in the two provided types are exactly and perfectly achieved.

The specific geometry provided in the two prescriptive cross-sectional types are based on the need for a secure grasp. The maximum perimeter measurements on Type I handrails and the "finger recess" on Type II handrails are intended to allow the user's hand to wrap around four or more planes (sides) of the handrail, creating a "hook" effect in the grasping. A larger perimeter or a lack of a finger recess may leave only three sides of the handrail within reach of the stair user's fingers, creating a weaker "pinch" of the rail rather than a "grasp." For example: The grasp on a handrail should be similar to a gymnast's grasp of a high bar, where fingers can wrap around the bar, as compared to trying to hang yourself from the underside of the deck joists by only pinching the two sides. The wrap of the fingers creates that grasp and is the intent of this section.

Field-fabricated handrails of conventional materials will often fit the parameters of Type I handrails, while Type II handrails are primarily seen in manufactured handrail systems (see Example 7-10). However, there are many manufactured products marketed for use as handrails whose cross-sectional shape does not fit within the two types detailed in this section. The ability to grasp these handrails would need to be considered in accordance with the intent of this section and with the approval of the local building official.

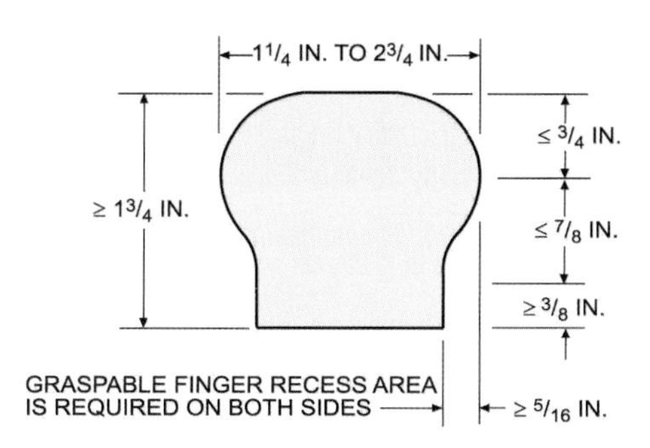

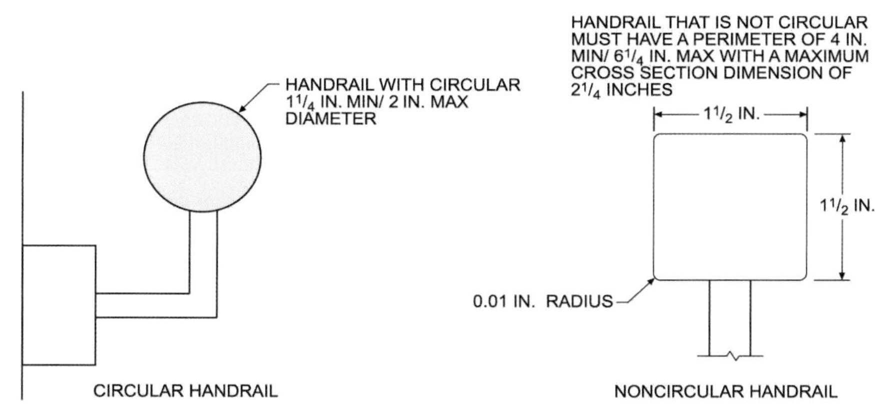

Example 7-10: These illustrations detail the prescriptive geometry allowed by the IRC to provide a graspable handrail. Other profiles would require approval from the local building official.

R311.7.7.4 Exterior wood/plastic composite handrails.

Wood/plastic composite handrails shall comply with the provisions of Section R317.4.

Discussion: See Chapter 5 for a discussion of Section R317.4.

Structural Design

Handrail live load design criterion is discussed in Part 2, Guards, under the discussion of Section R301.5.

Part Two: Guards

R312.1 Where required. *Guards* shall be located along open-sided walking surfaces, including stairs, ramps and landings, that are located more than 30 inches (762 mm) measured vertically to the floor or *grade* below at any point within 36 inches (914 mm) horizontally to the edge of the open side. Insect screening shall not be considered as a *guard.*

Discussion: The purpose of a guard is well explained within its IRC definition provided at the beginning of this chapter: "...minimizes the possibility of a fall from the walking surface to a lower level." This section clarifies that a "walking surface" includes surfaces such as stair treads and ramps and provides geometric parameters of when guards are required. Essentially, any building component at the edge of a raised walking surface could be defined as a guard, but whether it is required or not is dependent upon this section. Historically, a walking surface elevated 30 inches (762 mm) or more above an adjacent surface has required a guard. This height has typically been measured vertically from the edge of the surface to the floor or grade below. However, measuring in this manner may not be a true evaluation of the possibility for hazard. Often a deck may be built at the edge of a hill or retaining wall, where only the grade vertically below the edge of the deck may be less than 30 inches (762 mm), yet only a few feet away may be much higher (see Example 7-11). New to the 2009 IRC is a requirement for the height above grade to be measured vertically from a point 36 inches (914 mm) horizontally away from the edge of the deck. This provision provides increased safety to the guard requirement from a more realistic expectation of where someone would land if falling from a deck—not right at the edge, but a few feet out (see Example 7-12).

Example 7-11: This deck, while practically resting on the ground, is at the edge of a steep slope. A fall over the edge would likely send someone down the hill rather than to the ground immediately adjacent the deck.

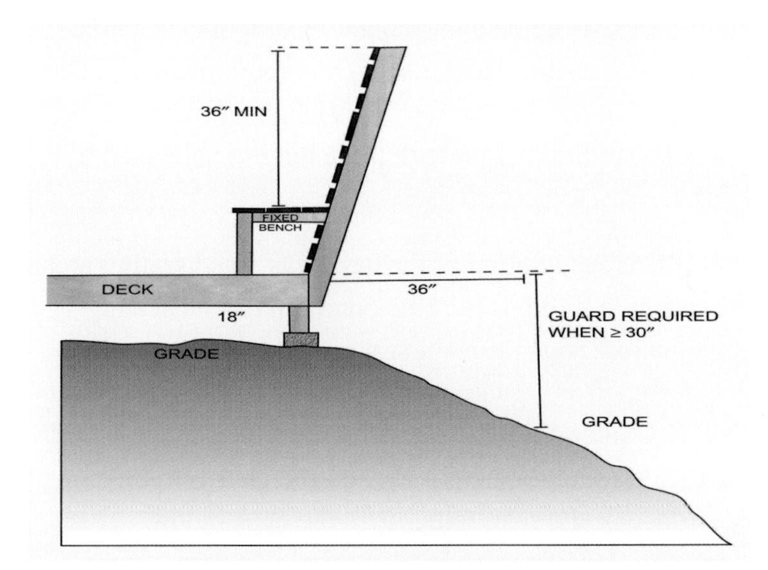

Example 7-12: Measuring 36 inches (914 mm) horizontally from the edge of the deck for determining the height above grade is consistent with the minimum size of landings for stairs and doors. This distance ensures that if a guard was not required, there would be a "landing" less than 30 inches (762 mm) below the deck. As discussed later in this chapter, when a fixed bench is adjacent a guard, the guard height must be measured from the bench surface.

R312.2 Height. Required *guards* at open-sided walking surfaces, including stairs, porches, balconies or landings, shall be not less than 36 inches (914 mm) high measured vertically above the adjacent walking surface, adjacent fixed seating or the line connecting the leading edges of the treads.

Exceptions:

1. *Guards* on the open sides of stairs shall have a height not less than 34 inches (864 mm) measured vertically from a line connecting the leading edges of the treads.
2. Where the top of the *guard* also serves as a handrail on the open sides of stairs, the top of the *guard* shall not be not less than 34 inches (864 mm) and not more than 38 inches (965 mm) measured vertically from a line connecting the leading edges of the treads.

Discussion: When a guard is required, based on the limitations in Section R312.1, it must be a minimum of 36 inches (914 mm) high, measured vertically from the walking surface it serves. This height is derived as a slight reduction from the 42-inch (1067 mm) average center of gravity of humans. The 6-inch (152 mm) reduction from this logically determined height was a compromise in private residential construction to allow more visibility over the guard from a seated occupant and to allow for more personal design freedom in private dwellings. Stairways are also allowed a reduction below the average center of gravity due to the common practice of using the stair guard as a handrail. A compromise is made between these components where guards are allowed to be as low as 34 inches (864 mm), measured vertically above the nosing of the stair treads. This height corresponds with the lowest allowable handrail height within the range established by Section R311.7.7.1.

New to the 2009 IRC is recognition of the hazard of fixed seating areas placed adjacent to a guard. It has become common to find seating incorporated into guard design, either a seat built against a guard, or a seat with a back designed as a guard (see Example 7-13). New provisions in the 2009 IRC intend to provide increased safety to these built-in seating areas by requiring the height of required guards to be measured from the surface of the seat, rather than the deck below. This rather significant code modification will essentially require a guard adjacent an 18-inch (457 mm) high seat to be constructed a total of 54 inches (1372 mm) above the deck below, such that it is the minimum 36-inch (914 mm) height above the seat surface.

What should be noted is that this modification does not change when guards are required, but rather how high they need to be. For example, a deck 29 inches (737 mm) above grade, with an 18-inch (457 mm) high seat built at the edge would not require any guards at all. A fall from this seat would result in a 47-inch (1194 mm) drop, well above the height for a required guard, but because "fixed seating" is not mentioned in Section R312.1, it would not apply. However, if the deck was 30 inches (762 mm) above grade, the guards would be required, and their height would then be measured from the adjacent bench seat (see Example 7-12).

Example 7-13: Without the bench installed, this guard is 36 inches (914 mm) high and is a compliant installation. When an 18-inch (457 mm) high bench is installed in a fixed location against the guard, the provisions in the IRC will now require the minimum guard height to be measured from the surface of the bench. The code is not retroactive, and an existing guard would not need to be increased in height with the code change.

R312.3 Opening limitations. Required *guards* shall not have openings from the walking surface to the required *guard* height which allow passage of a sphere 4 inches (102 mm) in diameter.

Exceptions:

1. The triangular openings at the open side of a stair, formed by the riser, tread and bottom rail of a *guard*, shall not allow passage of a sphere 6 inches (153 mm) in diameter.
2. *Guards* on the open sides of stairs shall not have openings which allow passage of a sphere $4^{3}/_{8}$ inches (111 mm) in diameter.

Discussion: The first word of this section, "required," is often overlooked, but is a very important adjective for complete comprehension of this section. Only guards that are required by Section R312.1 are subject to the opening and height restrictions. When a guard-like architectural element is installed where a complying guard is not required, then no provisions in this section apply (see Example 7-14).

For required guards, there are three different opening restrictions specific to certain locations, and all are intended to prohibit children from passing through a guard as compared to passing over a guard. Required guards located at the edge of flat walking surfaces, such as decks, floors, patios or even ramps, must be constructed in a manner such that a 4-inch (102 mm) diameter sphere cannot pass through the guard anywhere between the deck level and the 36-inch (914 mm) required height. Portions of guards that may extend higher than the minimum 36-inch (914 mm) required height are not regulated by this section and may have any size opening desired, as they are no longer at a height susceptible to child access.

Required guards at the sides of stairs are allowed a little more flexibility in their construction as they must only restrict passage of a 4 $^{3}/_{8}$-inch (111 mm) diameter sphere. This increase in opening size was provided by the IRC primarily to accommodate typical interior stair guards constructed of individual slender balusters, and allows for spacing of two balusters to be located uniformly on each tread, rather than the considerably decreased spacing of three balusters per tread. However, there is no language that would prohibit this increase in guard opening restrictions for any other stair, as it is recognized that the likelihood of children passing through guards at the sides of stairs is far less than that of level walking surfaces.

In the event that side stringers of stairs are notched in the shape of the riser height and tread depth, the triangular opening created within this space may be large enough that a 6-inch (152 mm) sphere cannot pass through (see Example 7-15)

Example 7-14: The height of this deck does not require guards; therefore, the design of the guard is not required to comply with the opening limitations.

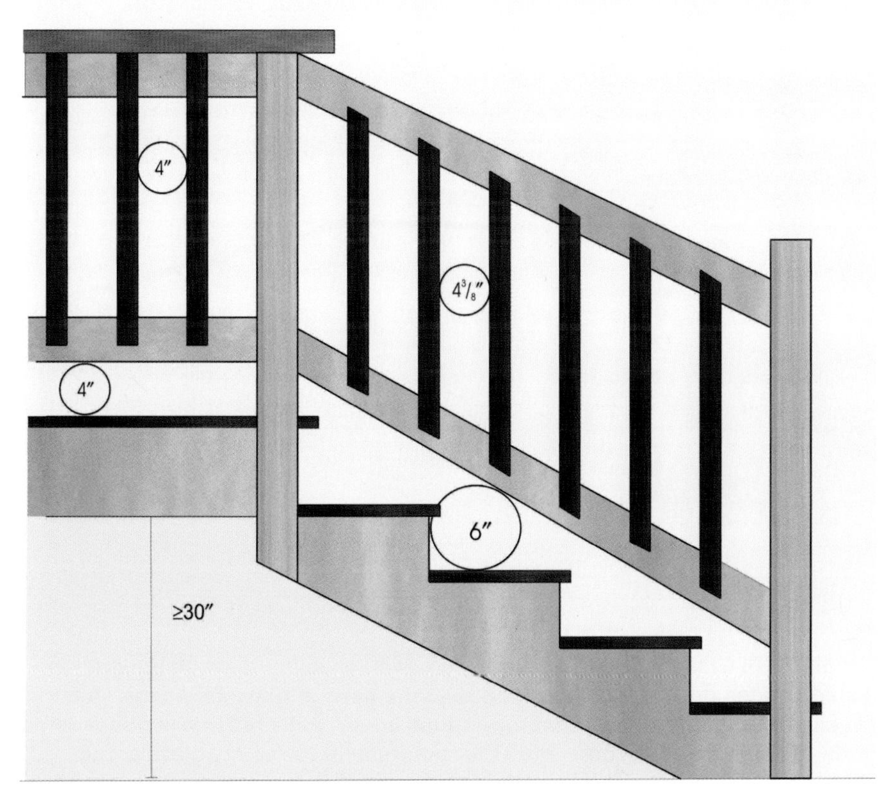

Example 7-15: This illustration details the three variations of opening restriction for guards as dependent on their location.

R308.4 Hazardous locations. The following shall be considered specific hazardous locations for the purposes of glazing:

4. All glazing in railings regardless of an area or height above a walking surface. Included are structural baluster panels and nonstructural infill panels.

Discussion: The organization of the IRC places this guard-related requirement in a more specifically related section, glazing. If your guard design incorporates any glass, referred to by the IRC as "glazing," it must be safety glass. Further details and requirements of safety glazing are provided in Chapter 2. This section clarifies that the size of the glazing and its height above the walking surface do not matter when used in a "railing" (guard). All glazed components in railings (guards) are regulated by this section with no exceptions, be it custom glass panels or manufactured glass balusters (see Example 7-16). The most common type of safety glazing is tempered glass. Section R301.5, which details the live load requirements of various building components, provides an additional requirement in the structural capabilities of glass guard infill and is covered in the following discussion.

Example 7-16: Glass infill provides an increased visibility through the guards, yet also creates potential for hazard due to the ease of glass breakage. A fall from any failed guard component may cause injury, but a fall that includes large shards of glass may be worse, thus the requirement for safety glazing and increased load capacity.

R301.5 Live load. The minimum uniformly distributed live load shall be as provided in Table R301.5.

TABLE R301.5
MINIMUM UNIFORMLY DISTRIBUTED LIVE LOADS
(in pounds per square foot)

USE	LIVE LOAD
Attics without storage[b]	10
Attics with limited storage[b,g]	20
Habitable attics and attics served with fixed stairs	30
Balconies (exterior) and decks[e]	40
Fire escapes	40
Guardrails and handrails[d]	200[h]
Guardrail in-fill components[f]	50[h]
Passenger vehicle garages[a]	50[a]
Rooms other than sleeping room	40
Sleeping rooms	30
Stairs	40[c]

For SI: 1 pound per square foot = 0.0479 kPa, 1 square inch = 645 mm^2,
1 pound = 4.45 N.

a. Elevated garage floors shall be capable of supporting a 2,000-pound load applied over a 20-square-inch area.
b. Attics without storage are those where the maximum clear height between joist and rafter is less than 42 inches, or where there are not two or more adjacent trusses with the same web configuration capable of containing a rectangle 42 inches high by 2 feet wide, or greater, located within the plane of the truss. For attics without storage, this live load need not be assumed to act concurrently with any other live load requirements.
c. Individual stair treads shall be designed for the uniformly distributed live load or a 300-pound concentrated load acting over an area of 4 square inches, whichever produces the greater stresses.
d. A single concentrated load applied in any direction at any point along the top.
e. See Section R502.2.2 for decks attached to exterior walls.
f. Guard in-fill components (all those except the handrail), balusters and panel fillers shall be designed to withstand a horizontally applied normal load of 50 pounds on an area equal to 1 square foot. This load need not be assumed to act concurrently with any other live load requirement.
g. For attics with limited storage and constructed with trusses, this live load need be applied only to those portions of the bottom chord where there are two or more adjacent trusses with the same web configuration capable of containing a rectangle 42 inches high or greater by 2 feet wide or greater, located within the plane of the truss. The rectangle shall fit between the top of the bottom chord and the bottom of any other truss member, provided that each of the following criteria is met.
1. The attic area is accessible by a pull-down stairway or framed in accordance with Section R807.1.
2. The truss has a bottom chord pitch less than 2:12.
3. Required insulation depth is less than the bottom chord member depth.

The bottom chords of trusses meeting the above criteria for limited storage shall be designed for the greater of the actual imposed dead load or 10 psf, uniformly distributed over the entire span.
h. Glazing used in handrail assemblies and guards shall be designed with a safety factor of 4. The safety factor shall be applied to each of the concentrated loads applied to the top of the rail, and to the load on the in-fill components. These loads shall be determined independent of one another, and loads are assumed not to occur with any other live load.

Discussion: Similar to the safety glazing requirement, this fundamental structural requirement for guard and handrail construction is organized into a different section of the IRC than that of handrails and guards. Table R301.5 describes the live load design criteria for different portions of a home. Included in this list are guards and handrails, and guard infill components. The table requires guards and handrails to be capable of resisting a 200 pound (890 N) live load. However, note d of this table cannot be overlooked as it describes the details of how that load is to be applied. Evaluation of the load-resistance capabilities of the guard or handrail assembly is only required to be based on a single concentrated 200 pound (890 N) force applied anywhere along the top of a guard or along the length of a handrail. This force must be resisted no matter the direction in which it has been applied. It is essential to understand that this single load must be resisted at all locations along the top of the guard or the handrail, but not at the same time.

Guard infill is composed of the all the remaining portions of the guard assembly beneath the top. Generally, most forces applied to guards will be at the top of the assembly, where people lean and sit. The infill portions of the guard requires a live load resistance of 50 pounds (223 N) distributed over a 1 square foot (0.0929 m^2) area, and similar to the 200 pound (890 N) load previously discussed, the details are in the footnote. The infill 50 pound (223 N) load resistance is to be evaluated assuming a single load of 50 pounds (223 N) applied in a horizontal direction to an area of 1 square foot (0.0929 m^2) on the infill material. The guard assembly as a whole does not have to be designed to resist both the infill load and the concentrated top-of-guard load simultaneously. For most guards designed using rails and balusters, it is important to keep the 1 square foot (0.0929 m^2) area in mind. Considering the common guard assembly of vertical balusters spaced at $3^3/_4$ inches (95 mm) apart, a single baluster is not required to resist this load—it will almost always be shared with others (see Example 7-17).

Note i addresses the structural analysis of glazing used in guard assemblies or assemblies used for the support of handrails. Not only must the glazing be safety glass (see R308.4, Item 4, in the previous discussion), but the load resistance required of both the top of the guard and the infill must be determined using a safety factor of four. Appropriately a job for a registered design professional or glazing professional, this safety factor will provide added insurance against the multiple hazards involved with failure of glass in these particular applications.

Often for architectural reasons, guards or handrails that are not required by the IRC are still installed. These extra features, whether required or not, should still be capable of resisting the above mentioned loads. A deck occupant cannot be expected to know which "guard-like" assemblies are strong enough to use, based on whether they were required or not. If it is available for an occupant to use as a guard, it should be structurally sound. This is intentionally different from the guard opening restrictions and guard height, previously discussed, which only apply to "required" guards.

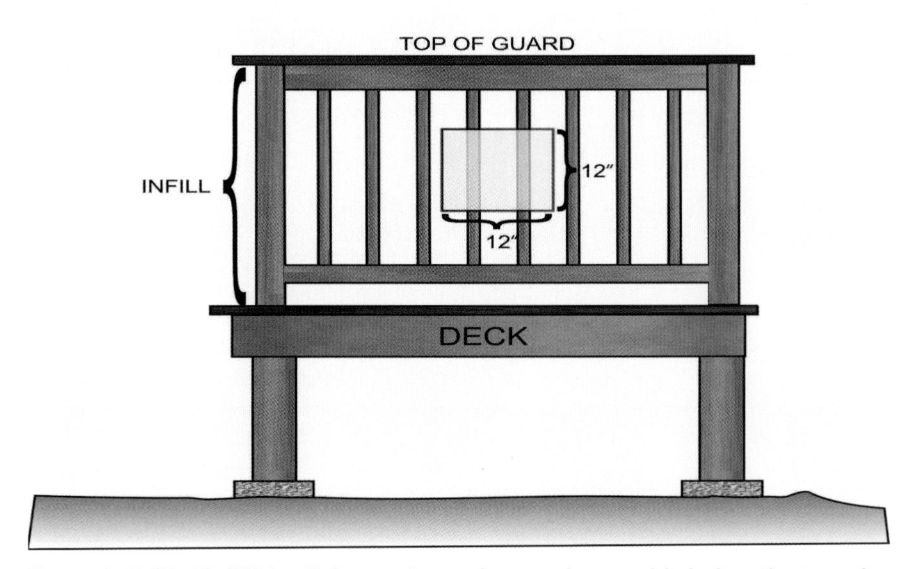

Example 7-17: "Infill" is all the portions of a guard assembly below the top edge. In this graphic of a typical deck guard of rails and balusters, at least two balusters will resist the 50 pound (223 N) horizontal load required by Table R301.5, as it is to be applied to a 1-foot (305 mm) area.

Side Note: Most manufactured guardrail systems should be manufactured to meet the IRC live load requirements, but it is the responsibility of the designer, builder and code official to verify this information. These systems also need to be installed exactly according to those manufacturer's written installation instructions. Field-constructed guards of typical dimensional lumber are not as simple to construct with certainty that the live load requirement will be met. Testing at prominent university laboratories has yielded results that indicate the typical installation of notched 4 by 4 guard posts may not be properly designed to resist the prescribed loads for the installation. The code does not provide prescriptive detailing for these post connections, and they may need to have calculations performed by a competent individual.

Chapter 8: Amenities

Introduction

Today's outdoor living contractors have helped create an industry where a deck is more than just a flat, exterior walking surface, but rather a multifunctional extension of the home. Often, a deck design will include an assortment of built-in amenities and features that require experience and knowledge beyond just that of joists and beams. For many of these amenities, the IRC provisions are nearly as extensive and involved as for the deck construction itself, and are more suited for a specific professional to handle. However, a basic understanding of IRC provisions related to these amenities can be very helpful in properly designing a deck to accommodate them. A deck contractor doesn't need to know how to install the wiring for a hot tub, as that is the licensed electrician's job, but knowing where the disconnect switches and other electrical components can be located may have a major effect on the design and installation. Many of the amenities discussed in this chapter will have additional IRC provisions related to site-specific aspects of their installation; this chapter does not intend to be all-inclusive. It does not provide all the answers to every installation, but should at least have you asking all the right questions.

Part One: Hot Tubs and Spas

Discussion: It is very common to see a deck designed to incorporate a hot tub or spa, be it in-ground, on the deck or adjacent to the deck. Whether the hot tub is to be installed immediately or is planned as a future installation, attention must be given to an assortment of building and electrical provisions. Most of these provisions cannot be left as a last-minute thought; as they can have dramatic affects on the shape of the deck, sizing of the structural members, location of glass, switches, receptacles and lights, and the presence of a security fence or safety cover. The requirements for a compliant hot tub installation cannot be left entirely to the expertise of the electrician; the correct planning must first occur.

R301.4 Dead load. The actual weights of materials and construction shall be used for determining dead load with consideration for the dead load of fixed service *equipment.*

Discussion: All engineered joist and beam span tables, either in the IRC or other sources, assume a uniformly distributed dead load, most often 10 psf (0.48 kPa) or 20 psf (0.96 kPa). The installation of a hot tub, or any other permanent, heavy amenity, must be made with consideration to how the structure will transmit the weight of it (dead load) to the supporting soil (see Example 8-1). All the members involved with supporting the actual weight of a hot tub filled with water must be evaluated during the design phase of the project. An 8-foot square (0.74 m^2) hot tub can easily weigh more than 1,000 pounds (454 kg), and distributed over a 64 square-foot (5.95 m^2) area yields an additional 16 pounds of dead load per square foot (0.77 kPa). These estimates are conservative, as many hot tubs exceed that weight. When analyzing the dead load, it is not necessary to include the weight of the occupants expected to use the tub, as their weight has been considered in the live load resistance of 40 psf (1.92 kPa).

Example 8-1: Heavy and permanent appliances, such as this outdoor kitchen and all hot tubs, must be considered as additional concentrated dead load during the design phase of the project.

R308.4 Hazardous locations. The following shall be considered specific hazardous locations for the purposes of glazing:

5. Glazing in enclosures for or walls facing hot tubs, whirlpools, saunas, steam rooms, bathtubs and showers where the bottom exposed edge of the glazing is less than 60 inches (1524 mm) measured vertically above any standing or walking surface.

   Exception: Glazing that is more than 60 inches (1524 mm), measured horizontally and in a straight line, from the water's edge of a hot tub, whirlpool or bathtub.

6. Glazing in walls and fences adjacent to indoor and outdoor swimming pools, hot tubs and spas where the bottom edge of the glazing is less than 60 inches (1524 mm) above a walking surface and within 60 inches (1524 mm), measured horizontally and in a straight line, of the water's edge. This shall apply to single glazing and all panes in multiple glazing.

Discussion: The area surrounding hot tubs, spas and swimming pools is usually wet. Even without children splashing, merely exiting one of these features will leave a trail of water at the perimeter. Mix this water with a walking surface, bare feet and an off-balanced occupant who has just exited the tub, and the perfect conditions for a slip and fall will exist. The IRC cannot prohibit a slip from sending someone into nearby glass, but it does minimize the hazard of impacting the glass through the requirement for safety glazing.

Item 6 under Section R308.4 describes this condition and the parameters of when it is a "hazardous location" in regard to glazing. All glazing (glass) within 60 inches (1524 mm) horizontally of the edge of the water, not the tub edge, and within 60 inches vertically of a walking surface below, must be safety glazed (see Example 8-2). The difficulty in applying this section comes with the question, "What is a walking surface?" It is not the intent of the IRC for all surfaces to be considered walking surfaces, rather those surfaces designed and constructed for the specific purpose of walking. Patio pavers, flagstone, concrete or a composite deck surrounding a hot tub are surfaces designed and intended specifically for the purpose of walking. This section should not be applied to every imaginable surface able to be walked upon, such as grass, river rock, dirt or other types of landscaped surfaces.

Item 5 intends to regulate glazing that "encloses" one of these water features. In this location, the concern arises from the movement of people within a hot tub, whirlpool, sauna or shower. Glazing at the periphery of a hot tub, within 60 inches (1524 mm) horizontally, would need to be safety glazed if within 60 inches (1524 mm) vertically from the "walking or standing surface" within the hot tub (see Example 8-3). Built-in seats or lounges molded in the inside of the hot tub shell may be intended for sitting, but are often used for standing during entry to the tub. When analyzing this safety glazing requirement, it is best to consider any flat horizontal surface within a hot tub as a "standing surface" whether intended or not.

Required labeling of safety glazing is described in detail in Chapter 2 of this book. Many times existing windows and doors may already be safety glazed due to features of the existing structure (see Example 8-4). Verification of the designation can confirm that existing glazing is safety glazing.

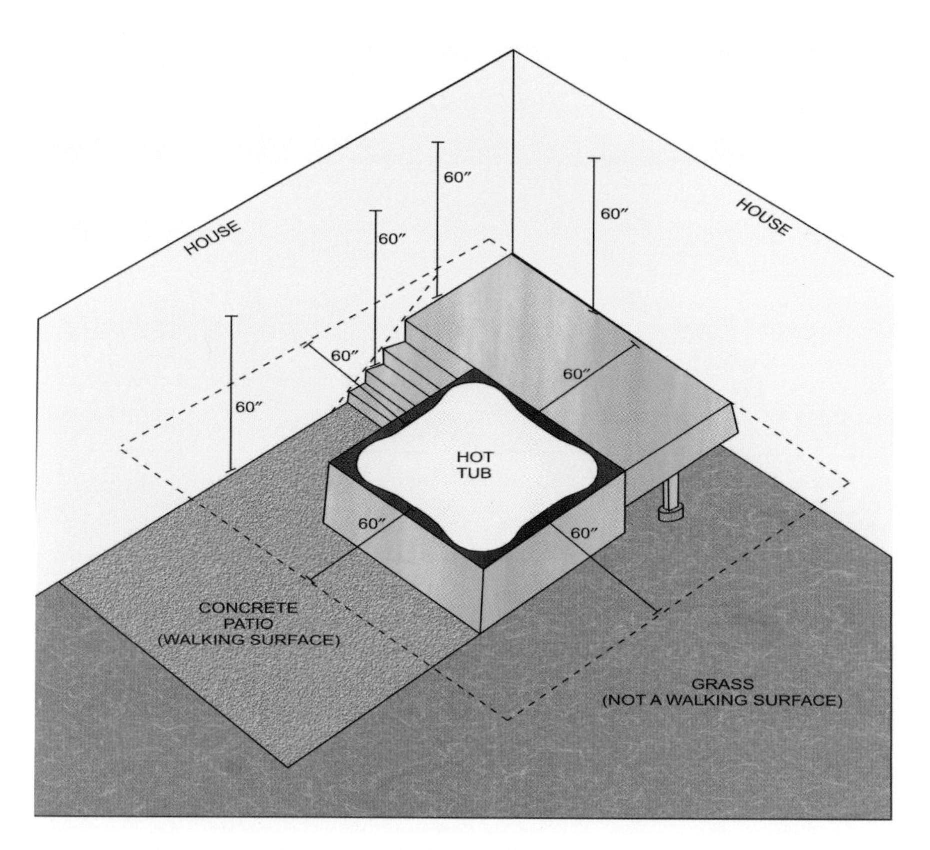

Example 8-2: Any glazing located within 60 inches (1524 mm) horizontally around the periphery of a hot tub may be in a hazardous location if its lowest edge is less than 60 inches (1524 mm) above the adjacent walking surface, such as the deck or concrete patio in this example.

Example 8-3: The two panes of glass acting as an enclosure to the side of the hot tub are required to be safety glazing in accordance with Item 5. The remaining two panes, beyond the boundaries of the hot tub, are also required to be safety glazing as they are within 60 inches (1524 mm) horizontally from the edge of the water.

Example 8-4: The sliding glass door in this photo should already be safety glazed, but the user should verify the label on the glass. Locating the hot tub near this door does not create any additional requirements for the door, but the light next to the door and the hot tub does not comply with Section E4203.4.3, provided following this discussion.

E4203.2 Switching devices. Switching devices shall be located not less than 5 feet (1524 mm) horizontally from the inside walls of pools, spas and hot tubs except where separated from the pool, spa or hot tub by a solid fence, wall, or other permanent barrier or the switches are listed for use within 5 feet (1524 mm). Switching devices located in a room or area containing a hydromassage bathtub shall be located in accordance with the general requirements of this code.

E4203.3 Disconnecting means. One or more means to simultaneously disconnect all ungrounded conductors for all utilization equipment, other than lighting, shall be provided. Each of such means shall be readily accessible and within sight from the equipment it serves and shall be located at least 5 feet (1524 mm) horizontally from the inside walls of a pool, spa, or hot tub unless separated from the open water by a permanently installed barrier that provides a 5 foot (1524 mm) or greater reach path. This horizontal distance shall be measured from the water's edge along the shortest path required to reach the disconnect.

E4203.4.1 Outdoor location. In outdoor pool, outdoor spas and outdoor hot tubs areas, luminaires, lighting outlets, and ceiling-suspended paddle fans shall not be installed over the pool or over the area extending 5 feet (1524 mm) horizontally from the inside walls of a pool except where no part of the luminaire or ceiling-suspended paddle fan is less than 12 feet (3658 mm) above the maximum water level.

E4203.4.3 Existing lighting outlets and luminaires.

Existing lighting outlets and luminaires that are located within 5 feet (1524 mm) horizontally from the inside walls of pools and outdoor spas and hot tubs shall be permitted to be located not less than 5 feet (1524 mm) vertically above the maximum water level, provided that such luminaires and outlets are rigidly attached to the existing structure and are protected by a ground-fault circuit-interrupter.

E4203.4.5 GFCI protection in adjacent areas. Luminaires and outlets that are installed in the area extending between 5 feet (1524 mm) and 10 feet (3048 mm) from the inside walls of pools and outdoor spas and hot tubs shall be protected by ground-fault circuit-interrupters except where such fixtures and outlets are installed not less than 5 feet (1524 mm) above the maximum water level and are rigidly attached to the structure.

Discussion: Presented to us as young children when we were rushed out of the local pool at the slightest threat of lightning, electricity and water make for a tremendous hazard. With the intention of keeping these two from meeting, the IRC provides strict provisions for any electrical hazards near these types of recreational bodies of water. All of the above sections are related to a 5-foot (1524 mm) horizontal distance, a distance that the average adult's arm length will not be able to reach.

Section E4203.3 requires a single means to disconnect all the ungrounded circuit conductors feeding equipment serving a hot tub, for the purpose of de-energizing the equipment for repair and maintenance purposes. This disconnect must be within sight of the hot tub, for reassurance to the maintenance personal that it will not be re-energized without their notice. As well as having an unobstructed view, "in sight" includes a maximum distance of 50 feet (15 240 mm). The disconnecting means and all other switches, as described in Section E4203.2, must maintain a 5-foot (1524 mm) horizontal distance from the inside edge of the hot tub, the water's edge (see Example 8-5). Reflecting the intent and purpose of this section, to be out of arm's reach, a permanent barrier can be installed between the hot tub and switches, including the disconnecting means. This barrier allows a switch to be less than the minimum 5-foot (1524 mm) horizontal distance, provided it is oriented such that a 5-foot (1524 mm) reach could not extend around the barrier and to the switch. Utilizing a barrier for a reduced distance to the disconnecting means requires a transparent barrier or other orientation that would still allow it to be seen from the hot tub location.

Lighting outlets, luminaries (light fixtures) and ceiling-mounted paddle fans also create an electrical hazard when installed near bodies of water. For new installations of electrical lighting and fan equipment, the same 5-foot (1524 mm) horizontal rule, previously described, also applies. However, this equipment can be installed within this horizontal distance and even above the hot tub, provided the lowest portion of the equipment is at least 12 feet (3658 mm) above the maximum level of the water, the flood-level edge of the hot tub. When existing lighting outlets or luminaries are within a 5-foot (1524 mm) horizontal distance from a new hot tub installation, the IRC allows a reduced vertical height from 12 feet (3658 mm) down to 5 feet (1524 mm). This reduction, however, requires the branch circuit to be protected by a GFCI (ground-fault circuit-interrupter) device. A GFCI device is designed for protection of people from electrical hazards.

Even when the lighting equipment is more than 5 feet (1524 mm) from the edge of the hot tub, the possibility for splashing or dripping water in areas around the tub can still create a hazard. Section E4203.4.5 requires this equipment, when between 5 and 10 feet (1524 mm and 3048 mm) horizontally from the water's edge, and when less than 5 vertical feet (1524 mm) from the maximum water level, to be provided with GFCI protection.

Example 8-5: The disconnecting means for this hot tub is installed at the edge of the deck at a distance just over 5 feet (1524 mm) from the inside edge of the tub and is "in sight" from the tub. This installation complies with IRC Section E4203.3.

E4204.2 Bonded parts. The parts of pools, spas, and hot tubs specified in Items 1 through 7 shall be bonded together using insulated, covered or bare solid copper conductors not smaller than 8 AWG or using rigid metal conduit of brass or other identified corrosion-resistant metal. An 8 AWG or larger solid copper bonding conductor provided to reduce voltage gradients in the pool, spa, or hot tub area shall not be required to be extended or attached to remote panelboards, service equipment, or electrodes. Connections shall be made by exothermic welding or by listed pressure connectors or clamps that are labeled as being suitable for the purpose and that are made of stainless steel, brass, copper or copper alloy. Connection devices or fittings that depend solely on solder shall not be used. Sheet metal screws shall not be used to connect bonding conductors or connection devices:

7. Metal wiring methods and equipment. Metal-sheathed cables and raceways, metal piping, and all fixed metal parts shall be bonded.
   Exceptions:
   1. Those separated from the pool by a permanent barrier shall not be required to be bonded.
   2. Those greater than 5 feet (1524 mm) horizontally from the inside walls of the pool shall not be required to be bonded.
   3. Those greater than 12 feet (3658 mm) measured vertically above the maximum water level of the pool, or as measured vertically above any observation stands, towers, or platforms, or any diving structures, shall not be required to be bonded.

Discussion: Metals of all types have one big thing in common: high electrical conductivity. As discussed previously, electrical equipment near bodies of water create quite a potential hazard. Electrical equipment, however, can also be damaged or otherwise misused such that it may energize other metal components nearby. These components may then become as hazardous as the equipment itself, and must be separated from bodies of water by the same horizontal 5-foot (1524 mm) distance previously discussed.

In lieu of separating these metal items from the hot tub by a distance of 5 feet (1524 mm), they can also be bonded together with one another, thus keeping all metal at an equal electrical potential. Bonding and grounding are outside the scope of this book and are respectfully the work of a licensed electrician. What is important and within the expectations of a deck design is the awareness of metal objects designed near a hot tub and the possibility that these objects will need to have a conductor bonding them to one another. Bonding requires the contact of metal to metal. For a bonding lug, the object would need to be drilled to access raw internal metal, and for a bonding clamp, all finish material would need to be removed at the clamp location for a metal to metal contact. This could result in removal of a corrosion-resistant finish or other such modifications that may not be ideal to the look of the finished project (see Example 8-6).

Example 8-6: The steel guard assemblies on two sides of this hot tub are within the horizontal distance requiring bonding of metal parts. Each guard section within the specified region would be required to be bonded to the metal parts of the hot tub.

E4203.6 Overhead conductor clearances. Except where installed with the clearances specified in Table E4203.5, the following parts of pools and outdoor spas and hot tubs shall not be placed under existing service-drop conductors or any other open overhead wiring; nor shall such wiring be installed above the following:

1. Pools and the areas extending 10 feet (3048 mm) horizontally from the inside of the walls of the pool;
2. Diving structures; or
3. Observation stands, towers, and platforms.

Overhead conductors of network-powered broadband communications systems shall comply with the provisions in Table E4203.5 for conductors operating at 0 to 750 volts to ground.

Utility-owned, -operated and -maintained communications conductors, community antenna system coaxial cables and the supporting messengers shall be permitted at a height of not less than 10 feet (3048 mm) above swimming and wading pools, diving structures, and observation stands, towers, and platforms.

TABLE E4203.5
OVERHEAD CONDUCTOR CLEARANCES

	INSULATED SUPPLY OR SERVICE DROP CABLES, 0-750 VOLTS TO GROUND, SUPPORTED ON AND CABLED TOGETHER WITH AN EFFECTIVELY GROUNDED BARE MESSENGER OR EFFECTIVELY GROUNDED NEUTRAL CONDUCTOR (feet)	ALL OTHER SUPPLY OR SERVICE DROP CONDUCTORS (feet)	
		Voltage to ground	
		0-15 kV	Greater than 15 to 50 kV
A. Clearance in any direction to the water level, edge of water surface, base of diving platform, or permanently-anchored raft	22.5	25	27
B. Clearance in any direction to the diving platform	14.5	17	18

For SI: 1 foot = 304.8 mm.

Discussion: Overhead service conductors cannot be installed above pools, including hot tubs, or above the area extending 10 feet (3048 mm) horizontally from the inside edge of the hot tub enclosure, unless at least $22^1/_2$ feet (6858 mm) above the surface of the water and up to 27 feet (8830 mm) above. Generally, these clearances cannot be easily met or verified, leaving placement of the hot tub 10 feet (3048 mm) or more horizontally from the overhead cables as the best option. Telephone and coaxial cables, which distribute only low-voltage currents, can be as low as 10 feet (3048 mm) above the maximum water level.

E4203.1.1 Location. Receptacles that provide power for water-pump motors or other loads directly related to the circulation and sanitation system shall be permitted to be located between 6 feet and 10 feet (1829 mm and 3048 mm) from the inside walls of pools and outdoor spas and hot tubs, and, where so located, shall be single and of the locking and grounding type and shall be protected by ground-fault circuit interrupters.

Other receptacles on the property shall be located not less than 6 feet (1829 mm) from the inside walls of pools and outdoor spas and hot tubs.

E4203.1.2 Where required. At least one 125-volt, 15- or 20-ampere receptacle supplied by a general-purpose branch circuit shall be located a minimum of 6 feet (1829 mm) from and not more than 20 feet (6096 mm) from the inside wall of pools and outdoor spas and hot tubs. This receptacle shall be located not more than 6 feet, 6 inches (1981 mm) above the floor, platform or grade level serving the pool, spa or hot tub.

E4203.1.3 GFCI protection. All 15- and 20-ampere, single phase, 125-volt receptacles located within 20 feet (6096 mm) of the inside walls of pools and outdoor spas and hot tubs shall be protected by a ground-fault circuit-interrupter. Outlets supplying pool pump motors from branch circuits with short-circuit and ground-fault protection rated 15 or 20 amperes, 125 volt or 240 volt, single phase, whether by receptacle or direct connection, shall be provided with ground-fault circuit-interrupter protection for personnel.

Discussion: Just as for lighting outlets and adjacent metal objects, receptacle outlets for cord-and-plug connected appliances also create an electrical hazard when located near bodies of water such as hot tubs. This type of electrical equipment has an additional concern, however; the attachment of a cord-and-plug connected appliance effectively extends the electrical hazard closer to the water by the distance of the appliance cord. Generally, most appliances have a 6-foot (1829 mm) cord, thus the increase in the horizontal clearance in these sections from 5 feet (1524 mm) to 6 feet (1829 mm).

At least one receptacle outlet must be provided between 6 feet (1829 mm) and 20 feet (6096 mm) from the inside edge of the hot tub enclosure (see Example 8-7). This required outlet is intended to provide power to service related equipment required by service personnel and must be no more than 6 feet and 6 inches (1981 mm) above the grade or walking surface below. All receptacle outlets within 20 feet (6096 mm) horizontally of the hot tub, regardless of their function or intended purpose, must be GFCI protected.

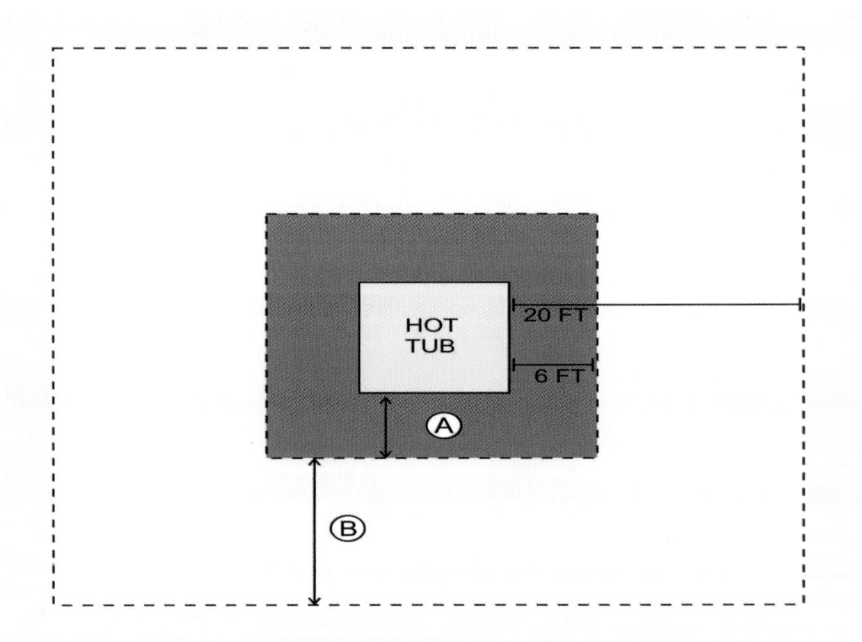

Example 8-7: Receptacle outlets are prohibited in the region labeled "A" in this example. In the shaded region, labeled as "B," one receptacle outlet is required and must be less than 6 feet and 6 inches (1981 mm) above grade. This outlet is required for service personnel, but any number of additional outlets may also be provided in this region. All receptacle outlets within the 20-foot (6096 mm) boundaries of region "B" must be protected by a ground-fault circuit-interrupter (GFCI), as well as all outdoor receptacle outlets.

Hot Tub Barrier (IRC Appendix)

Regardless of all the safety features that can be included in a hot tub installation, a child can drown in a mere inch of water, and certainly in a full hot tub. A security barrier is intended to prevent a child from gaining access to a hot tub. There are no security barrier provisions in the main text of the IRC, as many jurisdictions vary in this regulation. There is, however, an appendix chapter of the IRC for security barriers which can be independently adopted by the jurisdiction, and even when not adopted, provides a nationally accepted standard that is often emulated in locally developed swimming pool codes.

Appendix G, Swimming Pools, Spas and Hot Tubs

Extensive details regarding security barrier design have not been provided in this book, due to the commonly used exception related specifically to spas and hot tubs. Many jurisdictions may modify Appendix G when adopting.

AG105.5 Barrier exceptions. Spas or hot tubs with a safety cover which complies with ASTM F 1346, as listed in Section AG107, shall be exempt from the provisions of this appendix.

Discussion: The provisions for security barriers around swimming pools, hot tubs and similar recreational bathing features are considerably extensive and detailed, and are all related to prohibiting a child from finding a way over or through a barrier. Depending on the material and design of the barrier, protruding surfaces, chain-link fence openings, horizontal cross members and what side of the barrier they are located on are all considered in this section. Attention is also given to the closing and latching of gates, and to keeping them closed and latched, by regulating the location, height and access to the latch mechanism. Even the children who live at the property and are located inside the home must be provided protection from the water, in the form of door alarms or security barriers between the water and the house.

Conveniently, in regard to hot tubs and spas only, there is a simple exception to the barrier provisions; thus the extensive barrier requirements have not be provided in this book. If a security cover complying with ASTM F 1346 is provided, that is all that is required. Compliance with the provisions of this standard is verified by locating a label on the cover stating such compliance. ASTM F 1346 tests the ability of a cover to perform in a manner that will suitably protect a child less than five years of age from removing or otherwise passing through the cover. Child-resistant latching capabilities, ability to support 275 pounds (125 kg) (for weight of child and adult rescuer) and appropriate safety warnings are the primary features verified under compliance with this standard. When incorporating a hot tub into a deck design, it is strongly encouraged that a compliant safety cover be used. The additional cost for this type of cover will generally be far less than the design and construction of a fully compliant barrier perimeter, and far less inconvenient to the occupants than an alarm at their back door.

Part Two: Bench Seats

Bench seats are a common amenity often designed into new custom deck projects. The IRC does not provide any criteria for the construction of built-in seating itself, but rather how it may affect other requirements. When measuring the height of guards at the edge of walking surfaces greater than 30 inches (9144 mm) above the floor below, the presence of built-in seating at the guard location requires the minimum required height of the guard to be measured from the surface of the seat. More details of this IRC provision are discussed under Section R312.2 in Chapter 7 of this book.

When a bench is built at the edge of a deck and is acting as a required guard, it must be designed as a guard and capable of satisfying all the guard requirements in Chapter 7 of this book. For all other benches, it is suggested to construct them with the same load resistance required at the top of guards.

Part Three: Fuel-Gas Burning Equipment and Piping

Discussion: In creating an outdoor extension of a home, the warmth, ambiance and function of fireplaces, fire pits and barbeques have become commonplace. Whether providing a centerpiece for an outdoor living area or a kitchen for outdoor cooking, there is usually something in common, that being fuel gas. Appliances that burn gas for fuel, be it liquid petroleum gas (LPG) or natural gas, have inherent hazards in the delivery, combustion and exhausting of fuel gas. Although in many jurisdictions a state-licensed plumber is not required for gas pipe installations, it is highly recommended that an experienced professional be employed for the assembly of new gas distribution branches and appliance hookups. The following IRC sections are intended to provide a deck contractor the background information necessary for planning and designing a fuel-gas burning appliance within a deck, but not as a comprehensive installation guide.

G2408.1 (305.1) General. *Equipment* and *appliances* shall be installed as required by the terms of their approval, in accordance with the conditions of listing, the manufacturer's instructions and this *code*.

Manufacturers' installation instructions shall be available on the job site at the time of inspection. Where a *code* provision is less restrictive than the conditions of the listing of the *equipment* or *appliance* or the manufacturer's installation instructions, the conditions of the listing and the manufacturer's installation instructions shall apply.

Unlisted *appliances approved* in accordance with Section G2404.3 shall be limited to uses recommended by the manufacturer and shall be installed in accordance with the manufacturer's instructions, the provisions of this *code* and the requirements determined by the *code official*.

Discussion: The location and installation of fuel-gas burning appliances must be acceptable under the provisions of a number of documents, most of which are represented in the manufacturer's installation instructions. The approval from the building official is usually determined by the installation instructions, which are derived from the conditions of the appliance's listing. When designing one of these appliances into a deck project, selection of the appliance should be made prior to plan submittal. Likewise, the manufacturer's installation instructions must be on site for all the inspections, as many different conditions may require verification early in the construction process. Distance to combustible materials, venting methods and the magnitude of gas delivered must be evaluated for each appliance. Often this information comes from the manufacturer's installations instructions and is more restrictive than the provisions of the IRC; in those cases, the more restrictive shall govern (see the discussion of Section R102.4 in Chapter 1).

G2417.1.3 (406.1.3) New branches. Where new branches are installed to new *appliances*, only the newly installed branches shall be required to be *pressure tested.* Connections between the new *piping* and the existing *piping* shall be tested with a noncorrosive leak-detecting fluid or other *approved* leak-detecting methods.

Discussion: All new branches of gas piping shall be pressure tested prior to use for gas distribution. The new segment of assembled pipe and fittings must have a cap installed at the point of appliance connection with the appliance shut-off valve open. The other end, at the connection to the existing gas distribution system, must also be capped for the test. The pressure test must be verified by the local inspector, and hold pressure for a period of at least 10 minutes. Once the pressure test has been completed, the capped ends can be connected to the existing system and the appliance, where a liquid test can be performed. Do not subject a gas distribution system to a pressure test while it is still connected to appliances, as this may damage an appliance's internal gas valve and is specifically prohibited by the IRC.

G2415.9 (404.9) Protection against corrosion. Metallic *pipe* or *tubing* exposed to corrosive action, such as soil condition or moisture, shall be protected in an *approved* manner. Zinc coatings (galvanizing) shall not be deemed adequate protection for *gas piping* underground. Where dissimilar metals are joined underground, an insulating coupling or fitting shall be used. *Piping* shall not be laid in contact with cinders.

Discussion: When installing gas pipe in an exterior location, ferrous metal pipe (black iron) must be protected from corrosion. While this is generally achieved with the use of rust-inhibiting primer and paint, all methods must be "satisfactory" to the code official, otherwise referred to in the IRC as "approved." When using corrugated stainless steel tubing (CSST), installation requirements are specific to the manufacturer;

however, it is a general requirement that the outer jacket on the tubing remain intact or have proper repairs made to protect the steel tubing from unexpected corrosion. Black iron pipe and CSST cannot be installed underground unless properly sleeved or wrapped, sealed and vented, nor can exterior gas pipe be installed less than $3^1/_2$ inches (89 mm) above the ground surface. Specifics for these two requirements are provided in other IRC sections that are not included in this publication.

G2414.6 (403.6) Plastic pipe, tubing and fittings. *Plastic pipe, tubing* and fittings used to supply *fuel gas* shall conform to ASTM D 2513. *Pipe* shall be marked "Gas" and "ASTM D 2513."

Discussion: Plastic components of a gas distribution system can only be used in exterior, underground applications where they will be sufficiently protected from physical damage. Details of these components are outside of the scope of this book and are appropriately the work of an experienced plumbing or mechanical contractor.

G2415.14 (404.14) Location of outlets. The unthreaded portion of *piping outlets* shall extend not less than 1 inch (25 mm) through finished ceilings and walls and where extending through floors, outdoor patios and slabs, shall not be less than 2 inches (51 mm) above them. The *outlet* fitting or *piping* shall be securely supported. *Outlets* shall not be placed behind doors. Outlets shall be located in the room or space where the *appliance* is installed.

> **Exception:** Listed and labeled flush-mounted-type quick-disconnect devices and listed and labeled *gas convenience outlets* shall be installed in accordance with the manufacturer's installation instructions.

Discussion: The end of a gas distribution branch, where an appliance is to be connected, will typical extend through the finish material that may have been concealing the pipe, such as decking or a wall. To provide sufficient area for the grip of a pipe wrench to tighten the shutoff valve or appliance connector, an unthreaded portion of the pipe end must be extended through the finish material and be securely supported and restrained against movement. For pipe terminations through walls or ceilings there must be at least 1 inch (25 mm) of unthreaded pipe exposed, and for terminations through floors (decking) at least 2 inches

(51 mm) must be exposed (see Example 8-8). Unrelated to this provision, but in the same IRC section, the location of gas pipe terminations cannot be placed behind the swing of a door or in a space that is not intended for an appliance, as the IRC limits appliance connectors to a maximum length of 3 feet (914 mm).

Example 8-8: The black iron gas pipe that will be concealed by the decking material must be primed and painted, at least $3^1/_2$ inches (89 mm) above the ground, supported at proper intervals and at the termination, and extended sufficiently through the deck material. This termination should have had the required appliance shut-off installed prior to the cap in order to be part of the pressure test and for ease of appliance connection.

G2424.1 (415.1) Interval of support. *Piping* shall be supported at intervals not exceeding the spacing specified in Table G2424.1. Spacing of supports for CSST shall be in accordance with the CSST manufacturer's instructions.

TABLE G2424.1
SUPPORT OF PIPING

STEEL PIPE, NOMINAL SIZE OF PIPE (inches)	SPACING OF SUPPORTS (feet)	NOMINAL SIZE OF TUBING SMOOTH-WALL (inch O.D.)	SPACING OF SUPPORTS (feet)
$^1/_2$	6	$^1/_2$	4
$^3/_4$ or 1	8	$^5/_8$ or $^3/_4$	6
$1^1/_4$ or larger (horizontal)	10	$^7/_8$ or 1 (horizontal)	8
$1^1/_4$ or larger (vertical)	Every floor level	1 or larger (vertical)	Every floor level

For SI: 1 inch = 25.4 mm. 1 foot = 304.8 mm.

Discussion: Gas pipe must be sufficiently supported so as to maintain the seals of the fittings and to protect the pipe from physical damage. Support intervals are determined by the inherent strength of the materials based on type and diameter in accordance with the above table. CSST support intervals are determined by the manufacturer and provided in the manufacturer's installation instructions.

G2413.1 (402.1) General considerations. *Piping systems* shall be of such size and so installed as to provide a supply of gas sufficient to meet the maximum *demand* and supply gas to each *appliance* inlet at not less than the minimum supply pressure required by the *appliance*.

Discussion: When adding a new appliance to an existing gas distribution system, the entire system and all the connecting appliances must be evaluated. Installation of a new appliance without considering the impact it may have on the whole system may limit the gas delivery to an existing appliance. The combustion products from fuel-gas combustion can become considerably more hazardous and inefficient if the gas/air mix is unbalanced. In order to provide sufficient gas to each appliance during a period when all appliances are simultaneously burning, the diameter, length, pressure, pipe material of the system and the total Btu input of each appliance must be considered. This evaluation can be very difficult, if not infeasible, in an existing system if any portion is concealed. To bypass this evaluation and the possibility that part of the existing piping may need to be increased in diameter, a new distribution system can be connected directly to the gas meter. In this installation, the existing system is unaffected, and the new gas piping can be located entirely outside, thus limiting the need to expose the interior distribution system or damage the finish material.

Part Four: Outdoor Sinks

Discussion: In the extension of a home into the outdoor environment, an outdoor kitchen is often present. While this usually consists of a simple corner for a barbeque grill, it may also be an elaborate space designed with cabinets, countertops, grills, stoves and sometimes sinks. When providing a sink in an exterior location, freezing temperatures are about the only concern of the IRC that differs from interior sinks. Freeze protection or winterization of both the drain and the water supply must be addressed and are discussed in the following sections.

P3201.3 Trap setting and protection. Traps shall be set level with respect to their water seals and shall be protected from freezing. Trap seals shall be protected from siphonage, aspiration or back pressure by an *approved* system of venting (see Section P3101).

Discussion: All drains serving plumbing fixtures must include a water seal (trap) that will prohibit escape of sewer gas at the fixture location. In geographic locations subject to freezing temperatures, a fixture located outdoors must be provided with a means to protect a water trap from freezing, thus preventing damage to the drain and/or loss of the water seal. While there may be many methods to achieve freeze protection, there are no methods specifically provided in the IRC. In most applications, the use of insulation alone will not sufficiently provide freeze protection, but will only slow the heat loss from the water. A steady heat source is required to maintain temperature above freezing by offsetting the heat lost through the insulation. In some geographic locations freezing temperatures may be for short enough periods that the heat from the sewer system or from an adjacent conditioned space may migrate through a drain system and may be sufficient to eliminate freezing. In other geographic locations the use of a plug-in electric heat tape is commonly used to maintain a steady source of heat to an insulated drain.

P2603.6 Freezing. In localities having a winter design temperature of 32F (0C) or lower as shown in Table R301.2(1) of this code, a water, soil or waste pipe shall not be installed outside of a building, in exterior walls, in *attics* or crawl spaces, or in any other place subjected to freezing temperature unless adequate provision is made to protect it from freezing by insulation or heat or both. Water service pipe shall be installed not less than 12 inches (305 mm) deep and not less than 6 inches (152 mm) below the frost line.

P2903.10 Hose bibb. Hose bibbs subject to freezing, including the "frost-proof" type, shall be equipped with an accessible stop-and-waste-type valve inside the building so that they can be controlled and/or drained during cold periods.

> **Exception:** Frostproof hose bibbs installed such that the stem extends through the building insulation into an open heated or semi*conditioned space* need not be separately valved (see Figure P2903.10).

Discussion: Freeze protection of water supply pipe is generally a much larger concern than drain pipe due to the pressure and supply of water in the water distribution system. Freezing of a water supply pipe is much more likely to cause damage to the pipe material, as there is no air within the pipe to compress upon expanding of the water into ice. Water supply pipe exposed to freezing conditions must almost always be provided a source of heat similar to that of heat tape, as mentioned in the previous discussion.

In applications where an exterior sink is placed against and facing an exterior wall of the existing structure, another method can be used to provide water to the fixture. Just like a hose connection outside a house, a frost-proof hose bib, with a hot and cold connection, can be installed in the exterior wall of the existing structure, with a stem extending into a heated interior space, or a stop-and-waste valve inside for draining during cold-weather months.

Part Five: Low-Voltage Lighting

Discussion: The electrical chapters of the IRC do not contain provisions for installation of low-voltage wiring systems. However, Section E3401.2 ends in the following sentence: "Electrical systems, equipment or components not specifically covered in these chapters shall comply with the applicable provisions of the NFPA 70." NFPA 70, also referred to as the *National Electrical Code* (NEC), is the most recognized national standard in regard to electrical installations and is the basis of the requirements contained within the IRC electrical chapters. Article 411 of the 2008 NEC regulates the installation of lighting systems operating at 30 volts or less. Low-voltage systems are still capable of producing heat sufficient for combustion, and their installation must be evaluated with as much concern as other electrical equipment (see Example 8-9).

Within this NEC article, it is required that low-voltage lighting be listed as a system or as a system assembled of individually listed components, such as lights, transformers and wires. Listing of electrical components refers to equipment and materials that have been tested in accordance with all related national standards by an approved testing agency in an installation as required by the manufacturer. Listed construction components must be installed in accordance with the manufacturer's installation instructions, which reflect its listing, as that is the manner in which the equipment was tested.

Another provision of this NEC article applies to low-voltage lighting within 10 feet (3048 mm) horizontally from the edge of a body of water (hot tub, pool, fountain, etc.). When installed within this 10-foot (3048 mm) region, low-voltage lighting fixtures (luminaires) are regulated in the same manner as standard voltage equipment under Article 680 of the NEC. Under this article, the 10-foot (3048 mm) horizontal distance from the water's edge is reduced to 5 feet (1524 mm) for luminaires less than 10 feet (3048 mm) above the water's surface (see Example 8-10). There are other provisions of this article that must be addressed for low-voltage lighting less than 10 feet (3048 mm) from the water's edge, yet they are outside the scope of this publication, and are more suited for evaluation by a licensed electrician. Low-voltage electrical equipment that is listed for use near or in bodies of water is also permitted, provided the equipment is installed in accordance with its listing.

Example 8-9: The wire connections of this post-mounted low-voltage light were concealed beneath the top cap of the guard assembly. An installation that was not in accordance with the manufacturer's installation instructions resulted in arcing, overheating and ignition of the wood guard. Luckily, the occupant noticed the smoke lingering around the area and was able to extinguish the fire before any major damage occurred.

Example 8-10: The low-voltage lights in this photo provide the required illumination for this exterior stairway, as discussed in Chapter 6 of this book. However, these lights are installed within a 5-foot (1524 mm) horizontal distance to the edge of the water, and are only permitted if listed and labeled for installation within that proximity.

Part Six: Adding a Door from the House

Often in deck construction, a new exterior door from the existing dwelling is desired. The following IRC sections and discussions related to this topic are in regard to the location and interaction of the new building opening as related to the existing structure. The structural, water-resistant and energy-code-related requirements are outside the scope of this book. The following information is intended to provide guidance to the overall design of the door addition, not the installation.

G2427.8 (503.8) Venting system termination location.

The location of venting system terminations shall comply with the following (see Appendix C):

2. A *mechanical draft* venting system, excluding *direct*-vent *appliances*, shall terminate at least 4 feet (1219 mm) below, 4 feet (1219 mm) horizontally from, or 1 foot (305 mm) above any door, operable window, or gravity air inlet into any building. The bottom of the vent terminal shall be located at least 12 inches (305 mm) above finished ground level.
3. The vent terminal of a *direct*-vent *appliance* with an input of 10,000 *Btu* per hour (3 kW) or less shall be located at least 6 inches (152 mm) from any air opening into a building, and such an *appliance* with an input over 10,000 *Btu* per hour (3 kW) but not over 50,000 *Btu* per hour (14.7 kW) shall be installed with a 9-inch (230 mm) vent termination *clearance*, and an *appliance* with an input over 50,000 *Btu/h* (14.7 kW) shall have at least a 12-inch (305 mm) vent termination *clearance*. The bottom of the vent terminal and the air intake shall be located at least 12 inches (305 mm) above finished ground level.

M1502.3 Duct termination. Exhaust ducts shall terminate on the outside of the building. Exhaust duct terminations shall be in accordance with the dryer manufacturer's installation instructions. If the manufacturer's installation instructions do not specify a termination location, the exhaust duct shall terminate not less than 3 feet (914 mm) in any direction from openings into buildings. Exhaust duct terminations shall be equipped with a backdraft damper. Screens shall not be installed at the duct termination.

P3103.5 Location of vent terminal. An open vent terminal from a drainage system shall not be located less than 4 feet (1219 mm) directly beneath any door, openable window, or other air intake opening of the building or of an adjacent building, nor shall any such vent terminal be within 10 feet (3048 mm) horizontally of such an opening unless it is at least 2 feet (610 mm) above the top of such opening.

Discussion: A door to a structure is considered an opening, and creates a pathway for air and air pollutants to migrate into a structure. The presence of bath or under-floor exhaust fans, dryer exhaust, range hood exhaust and fuel-burning equipment venting will often put a dwelling under negative pressure, thus drawing outside air into the interior. The wind pressures and atmospheric pressures around a structure can also create conditions where air is drawn into a dwelling. It is important to maintain a separation between the pollutants being exhausted from a home and the openings where they could be drawn back inside.

When adding a door to a home, the termination locations of all the pollutants associated with the home must be evaluated. The IRC sections above detail the clearance requirements of three common pollutants from building openings: products of fuel-gas combustion, clothes dryer exhaust and vents serving the sanitary drainage system. Bathroom exhaust and range hood exhaust are not considered pollutants and do not have any required clearances.

Section G2427.8 regulates clearances to building openings from fuel-gas burning vent terminations. The termination of a mechanical venting system must be a minimum of 4 feet (1219 mm) from a door or window, both beside or below the lowest part of the opening. If the vent termination is above the opening it must be at least 1 foot (305 mm) above. A mechanical vent system requires an appliance to have a fan that either induces a draft or forces the exhaust of the flue gasses inside the vent. These vent systems will typically be found in appliances such as furnaces, whereas many water heater and boilers utilize natural drafting. Natural draft appliances would not require these clearances; however, the clearances are still encouraged. Direct vent equipment, such as many fuel-gas fireplaces, require clearances based on the total Btu rating of the equipment, as detailed in Item 3 in Section G2427.8.

A dryer exhaust termination must be a minimum of 3 feet (914 mm) from any opening, measured in any direction, as detailed in Section M1502.2. Plumbing vent terminations, whether on a municipal sewer system or private sewage system, have the largest required clearances. These terminations must be at least 4 feet (1219 mm) below, 10 feet (3048 mm) horizontally or 2 feet (610 mm) above the closest edge of a building opening (see Example 8-11).

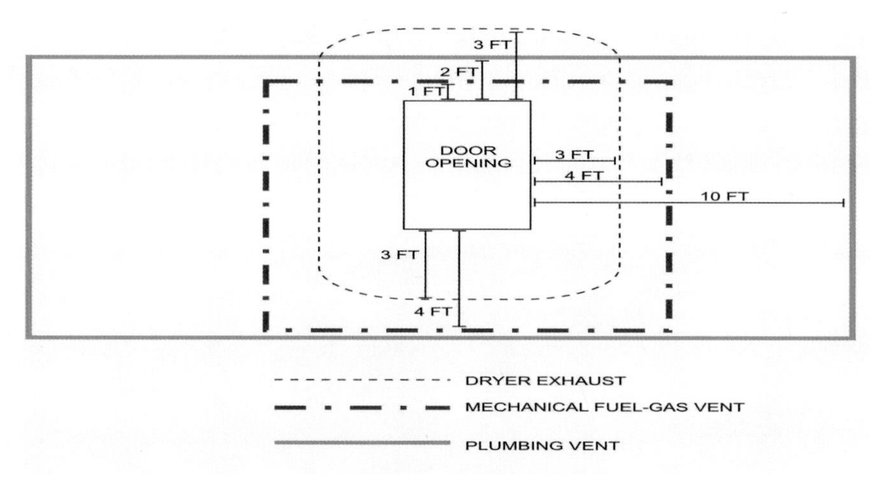

Example 8-11: The three clearances provided in the respective IRC sections published in this book are detailed in this example.

E3903.3 Additional locations. At least one wall-switch-controlled lighting outlet shall be installed in hallways, stairways, attached garages, and detached garages with electric power. At least one wall-switch-controlled lighting outlet shall be installed to provide illumination on the exterior side of each outdoor egress door having grade level access, including outdoor egress doors for attached garages and detached garages with electric power. A vehicle door in a garage shall not be considered as an outdoor egress door. Where one or more lighting outlets are installed for interior stairways, there shall be a wall switch at each floor level and landing level that includes an entryway to control the lighting outlets where the stairway between floor levels has six or more risers.

Exception: In hallways, stairways, and at outdoor egress doors, remote, central, or automatic control of lighting shall be permitted.

Discussion: The exterior door of a dwelling is usually the final interface between the privacy and security of an individual's home and the rest of the world outside. To maintain and provide occupant awareness of the world beyond their doors, the IRC requires exterior doors with grade level access to be provided illumination. The controls for this illumination must be located on a wall inside the dwelling. This IRC requirement must be satisfied when adding a new door to the exterior of a home, only when the door will lead to grade or to a walking surface leading to grade, such as a new deck. Similarly, and discussed under Section E3901.7 in Chapter 2 of this book, when a side of a dwelling is provided access to grade, that side must also be provided a with a grade-level accessible receptacle outlet.

Other than the addition of a new door, another condition may require evaluation of this IRC requirement. When a deck, previously without grade-level access, is modified with the addition of a stairway or guard opening leading to grade, the requirements for door illumination and a grade-level receptacle outlet would be applicable (see Example 8-12).

Example 8-12: If a new stairway was added to this deck such that it has grade-level access, a light must be installed on the exterior side of the door controlled by a switch at an interior wall location.

R308.4 Hazardous locations. The following shall be considered specific hazardous locations for the purposes of glazing:

2. Glazing, in an individual fixed or operable panel adjacent to a door where the nearest vertical edge is within a 24-inch (610 mm) arc of the door in a closed position and whose bottom edge is less than 60 inches (1524 mm) above the floor or walking surface.

   Exceptions:

   1. Decorative glazing.
   2. When there is an intervening wall or other permanent barrier between the door and the glazing.
   3. Glazing in walls on the latch side of and perpendicular to the plane of the door in a closed position.
   4. Glazing adjacent to a door where access through the door is to a closet or storage area 3 feet (914 mm) or less in depth.
   5. Glazing that is adjacent to the fixed panel of patio doors.

Discussion: Just as decks and stairs can create new hazardous locations where existing window glazing is present, so can the addition of a new doorway. Glass adjacent to a door can become a safety hazard due to the concentration of people moving nearby. A door is a feature in a wall that attracts people to and through it, and the concentration of moving people at these locations creates a greater opportunity for accidental contact with nearby glass. The swinging of a door and the possibility of panicking people evacuating a home also adds to the potential for accidents near doorways.

The IRC intends to limit the hazard of these accidents by requiring all glass within a 24-inch (610 mm) arc of either door jamb, and less than 60 inches (1524 mm) above a walking surface to be safety glazed. The edges of the door for which this arc is measured from are the operable portions only, and do not apply to any glazing that is adjacent to the fixed panel of a patio door (see Example 8-13).

Details are provided in the exceptions for this hazardous location with the intention of more specifically applying the section only to the actual hazards. Glazed openings in the door itself that cannot allow the passage of a 3-inch (76 mm) sphere are considered small enough that the safety glazing will not provide any significant additional protection. Similarly, decorative glass is already composed of small glass pieces held together in a solid border, and due to its artistic nature, is not usually constructed as safety glazing.

Two other exceptions are related to glazing locations that, while near doorways, are still considered less hazardous. If a permanent barrier exists between the glazing and the door, the occupant would be protected from contact with the glass. In a similar evaluation, glazing located in an assembly that is perpendicular to the door in a closed position is less likely to experience contact from people. However, due to the swinging of a door, only glazing perpendicular to the door on the latch side is exempted, as on the other side, the door itself may contact the glass.

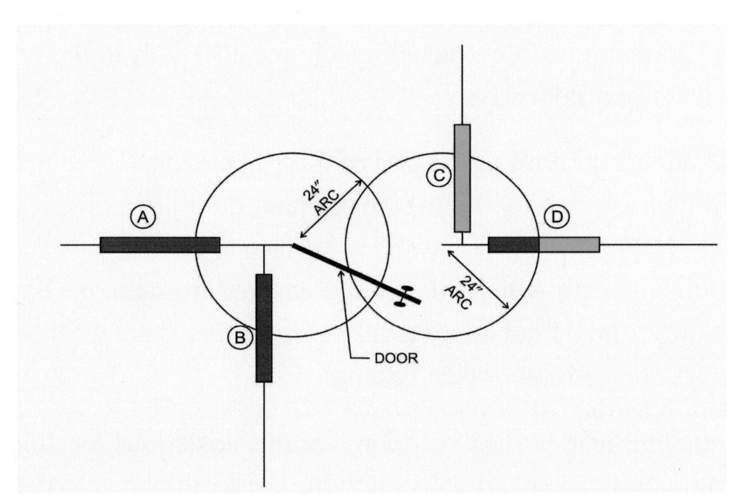

Example 8-13: This example displays various orientations of windows near a door. Window A and one of the panes in window D must be safety glazed due to their location within the 24-inch (610 mm) arcs measured at each side of the door jambs. The other pane in the window D assembly is outside of the arc, and is not required to be safety glazed, as each pane of glass is considered separately. Windows C and B are both perpendicular to the door in the closed position, but since window C is on the latch side of the door, only it is exempt from the safety glazing requirements.

Part Seven: Under-Deck Drainage Systems

There are many systems available in the market for transforming a deck surface into a water drainage system, and thus providing a dry area beneath a deck. These systems are generally considered as "storm drainage," similar to that of a gutter. The spaces created underneath these systems will not usually be evaluated in the same manner as other interior locations of the dwelling. Installation of electrical equipment under these drainage systems will generally be required as for wet locations, and treated as if they were outdoors and exposed to the weather. If, however, evidence was provided to the building official for evaluation of a deck drainage system as an "alternative" to the roof membrane requirements in the IRC, other uses beneath the deck may be approved.

Chapter 9: IBC Construction

Introduction

Decks serving the general public require considerably more evaluation in their design and construction than those built under the IRC for private residences and townhouses. Commercial decks usually serve many more people and many more functions. The people served are also much more unfamiliar with their surroundings than they would be at their own home, thus requiring increased safety provisions. The type of construction, occupancy type, occupancy load, means of egress system, presence of fire-resistant construction, fire-sprinkler systems and the necessity for accessible features for individuals with physical disabilities drive the need for a much deeper analysis of a proposed deck design.

This chapter is intended primarily as a general overview of the geometric variations of features in a commercial deck, as compared to those detailed in previous chapters of this book for residential decks. The issues presented above are outside the scope of this book and are respectfully the job of a design professional, followed by a review from the local building department. However, a brief discussion is provided as an introduction to some of the general fundamentals. In some states, state law may require a licensed architect or engineer for certain commercial projects.

Part One: Chapter One Variations

The administration provisions for the IBC are generally identical to those in the IRC, and are sometimes modified by the local jurisdiction prior to adoption. The IBC does not contain any provisions for other trades, such as plumbing, mechanical and electrical. These installations are regulated by other I-Code documents, such as the *International Plumbing Code®* (IPC®), *International Mechanical Code®* (IMC®), *International Fuel-Gas Code®* (IFGC®), *International Energy Conservation Code®* (IECC®), *National Electrical Code* (NEC) and related documents published by the National Fire Protection Association (NFPA). Working on commercial construction projects as a general contractor requires a much greater understanding of building codes and related codes than residential construction may require.

Unlike the IRC, the IBC does not have an exception for a required permit for decks less than 30 inches (762 mm) above grade. All decks constructed under the provisions of the IBC are required to be permitted.

Part Two: Chapter Two Variations

The interaction between deck construction and other systems in an existing structure, such as mechanical, plumbing and electrical systems and equipment, is a much greater concern in commercial construction. Higher voltage, greater fuel-gas consumption and delivery, larger scale exhaust and HVAC systems, and more vibrant fumes from a more frequently used plumbing system can all be present in commercial structures and pose a much greater risk to many more people than in private residential structures. There is also a possibility for many other features and systems to exist in a commercial building, ones that would need additional considerations. Fire-resistant construction, fire sprinkler systems, egress illumination and signage, and other systems serving specific occupancy-related uses are not present in private dwellings in IRC-regulated structures.

The open area around a building may also require evaluation prior to deck construction. When a building is designed, such as an apartment, the designer needs to consider provisions for open area around the building for fire department vehicle access. These types of conditions should be described in the building's original records and should be reviewed with the local building department during plan approval.

When greater numbers of people will be served by a commercial deck, or exiting across a deck, it will have an effect on the number of exits from the deck and possibly the illumination and exit signage of the deck space. A deck at a restaurant, for example, may be large enough that the number of people it can hold would require multiple exits. Even small issues, such as the direction of swing of a door from a building or from a gate in a deck guard is based on the number of occupants served and how they are intended to move between the spaces and to the exit. Generally, when a deck is large enough to serve 50 or more people, a design professional is strongly recommended, and may be required by state law, as this is a common threshold for dramatic variations in the allowable design.

Part Three: Chapter Three Variations

Foundation design is not significantly different in commercial decks than in residential decks, except in the case of tremendously large decks, where an engineer may be required. For decks added to existing commercial properties, the architect of record may also be involved in the design of any deck added to the structure. In the IRC there is a provision which allows free-standing decks which are not connected to the dwelling to be exempt from the frost depth requirement. This provision is not provided in the IBC, as all foundations supporting structures must be protected from the effects of frost heave, regardless of their connection, or lack thereof, to another structure.

Part Four: Chapter Four Variations

Residential decks need to be designed to support a 40 pound per square foot (1.92 kPa) live load, where commercial occupancies' live loads will vary based on the use. The designer needs to verify the required design loads for the project, as some of these commercial occupancies require more than the residential uses.

Part Five: Chapter Five Variations

Decking choices for commercial locations will be most affected by the type of construction of the existing building and location on the property in relation to the lot lines (fire-separation distance). The live load resistance required by the various commercial uses may also be much greater than the 40 pounds per square foot (1.92 kPa) required by the IRC. Reviewing the manufacturer's installation information or the product's testing results should yield this information for the deck designer.

Part Six: Chapter Six Variations

Stairways

Stairways in public locations must be constructed slightly different than in residential locations, due primarily to the increased number of people using them. To provide increased flexibility in private homes and to minimize the area of the interior floor space sacrificed for the stairway, residential stairways are allowed a reduction from the geometry required in commercial locations.

The tread depth of public stair treads can never be less than 11 inches (279 mm), and the maximum riser height cannot exceed 7 inches (178 mm). This geometry will result in a less-steep stairway than a residential counterpart, providing a greater ease of use and a lessened hazard. Unlike the IRC, public stairways have a minimum riser height of 4 inches (102 mm), intended to reduce the trip hazard of a small, hard-to-notice step. The landings at the top and bottom of a stairway also must be increased in length in the direction of travel. Landings at public stairways must be at least as deep in the direction of travel as the stairs are wide, but are never required to be more than 48 inches (1219 mm).

In stairways with winder treads, the minimum tread depth of 6 inches (152 mm) at the narrowest side for IRC structures is increased to a minimum of 10 inches (254 mm) for IBC structures. This increase will provide greater safety to the user, but also results in a reduced turn or curve of the overall stairway. Spiral stairways, as defined in Chapter 7 of this book, are much more regulated in public locations due to their narrow and difficult-to-use nature. These stairways can only be used within dwelling units or from a space not more than 250 square feet (23 m^2) with an occupant load of no more than five.

Multilevel decks with small changes of elevation between them are a common design practice in residential locations, as they effectively break up the space into various living areas and help to create the feel of an "outdoor home." In public locations, however, small changes of elevation add to the trip hazard from an unnoticed stairway, and generally decrease the ease of movement of large numbers of people. Generally, elevation changes of less than 12 inches (305 mm) between floor areas cannot be achieved by the use of a stairway, and a sloped surface is required. Smaller changes of elevation, 6 inches (152 mm) or less, must be equipped with either handrails or contrasting floor finishes.

There are exceptions to this rule in conditions where the elevation changes are at a door in certain occupancies, or are not required to be accessible.

Stairs in public locations, of any occupancy type, must be capable of resisting a 100 pound per square foot (4.8 kPa) uniformly distributed live load.

Stairway width for all occupancy types serving occupant loads of less than 50 is the same as what is required by the IRC for private stairs, 36 inches (914 mm). When the occupant load is 50 or greater, a stairway cannot be less than 44 inches (1118 mm) wide. Once the occupant load equals or exceeds this threshold value of 50, the total required egress width must be evaluated, and can sometimes require the stairway width to be greater than the minimum 44 inches (1118 mm). When considering the occupant load served by the stairway, it is not related to the occupant load of the deck alone, but all portions of the building designed to exit across the deck. A very small deck could require a very wide stairway if it is part of the exit path for a larger portion of the building. This evaluation is outside the scope of this book and is best suited for a design professional or the help from the local building department.

Unlike the IRC, illumination requirements are not specifically referenced for stairways or ramps, as all portions of the means of egress system (exit pathway) must be illuminated at all times a building is occupied. This includes all exterior portions of the building that serve to connect the occupants of a building to a public way. In most cases, illumination of deck or stair areas will already be provided at the exterior door location. However, large modifications to existing outdoor areas of public buildings or the creation of new outdoor spaces will likely require an evaluation of the illumination requirements as a whole.

Ramps

Ramps constructed on public decks are usually provided as a requirement for accessibility of disabled individuals. These ramps require much more consideration in their design than just the maximum slope, which is the same as in IRC structures, 1 unit vertically for each 12 units horizontally. Commercial ramps are limited in total height to a maximum of 30 inches (762 mm), at which point a landing is required. The landing must be at least 60 inches (1524 mm) deep in the direction of travel, and at least as wide as the widest ramp it serves. However, a landing between two ramps that change direction must be at least 60 inches by 60 inches (1524 mm by 1524 mm) square, as this provides sufficient turning space for an individual using a wheelchair (see Example 9-1).

Public ramps must also be provided with edge protection, intended to keep the wheel from a wheel chair or the end of a cane or walker from slipping over the edge of the ramp walking surface. Edge protection can be provided by a rail, curb or wall adjacent the ramp, in any manner or design desired provided it does not allow the passage of a 4-inch (102 mm) sphere where any portion of the sphere is within 4 inches (102 mm) of the ramp surface. As an alternative to the barrier, edge protection can be provided by the extension of the ramp surface horizontally at least 12 inches (305 mm) beyond the inside of the handrails (see Example 9-2)

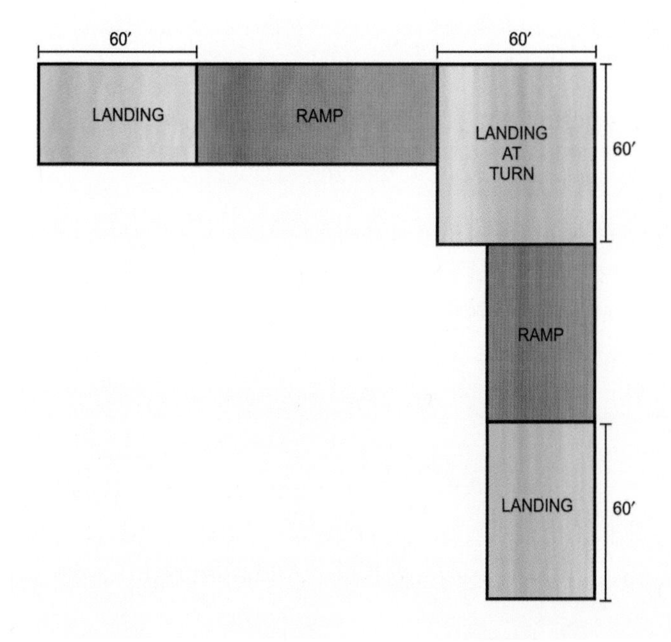

Example 9-1: Two different width ramps are detailed in this illustration. The landings at the end of each ramp must be at least as wide as the ramp served. The middle landing, where a turn is created, must be a minimum of 60 inches by 60 inches (1524 mm by 1524 mm), regardless of the width of the ramps.

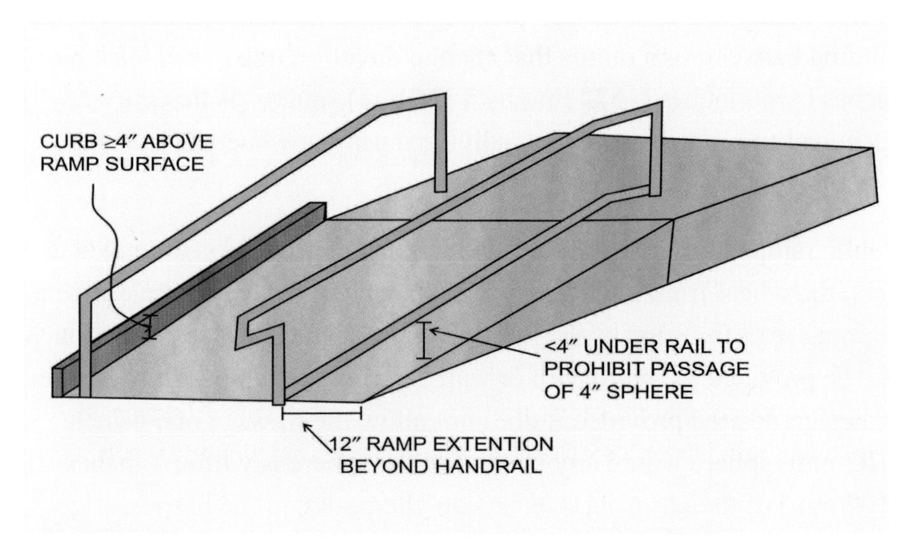

Example 9-2: Edge protection can be provided a number of ways. This illustration depicts the use of a ramp extension, curb and rail.

Part Seven: Chapter Seven Variations

Guards

As described in Chapter 7 of this book, the average center of balance of people is 42 inches (1067 mm) above the floor, thus the requirement for a minimum height of 42 inches (1067 mm) to the top of a required guard. At stairways, this height is not reduced to match the handrail height as it is in structures regulated by the IRC. At stairways that require a guard, a guard cannot function as the handrail, as is common in IRC structures. A separate handrail must be installed below the guard within the height range of 34 inches to 38 inches (864 mm to 965 mm), measured from the nosing of the stair treads, just as for IRC structures.

The opening restrictions for required guards in public locations also differ from the IRC, including required guards at the sides of stairs. The "4-inch-sphere" rule still applies, but only to a height of 36 inches (914 mm). Between 36 inches and 42 inches (914 mm and 1067 mm) the opening can be slightly larger, but not allowing the passage of a 4 $^{3}/_{8}$-inch (111 mm) sphere.

Handrails

Stairs constructed under the IBC, including a single riser between floors, must be provided handrails on both sides, regardless of their total number of risers. However, there are a number of exceptions under this IBC requirement. One exception, specific to decks and walkways, allows the omission of handrails when there is a single elevation change between landings, when the landings are greater in depth than the minimum required.

As explained in Part 6 of this chapter, stairway width varies based on the occupant load. When wide stairways are required due to a high occupant load, a different analysis of the required handrails is required. For these stairways one or more intermediate handrails within the stairway width may be required. The IBC requires intermediate handrails to be located such that a handrail is within 30 inches (762 mm) horizontally from any portion of the required stair width. Stairs that are constructed wider than required must have the handrails located in the area of the stairs which is in the most direct path of travel. This is the section of the stairs where the "required" width is considered to be located; the rest of the stairs are just extra and do not need a handrail within 30 inches (762 mm) (see Example 9-3).

Unlike handrails on IRC stairs, which do not need to extend beyond the first or last tread nosing, IBC handrails must have extensions. Extensions allow a stair user to grasp the handrail before traversing the steps. This provides an additional safety to stair users by providing a more noticeable invitation to use the provided handrails (see Example 9-4).

The live loads that guards and handrails must be capable of resisting are similar to that expected of private guards and handrails regulated by the IRC, but with an additional measure. Along with the concentrated load of 200 pounds (890 N) along the top and the infill load of 50 pounds over a 1 square foot (2.39 kPa) area, as discussed in Chapter 7 of this book, handrails and guards must also resist a 50 pound per linear foot (222 N) load applied in any direction along the top rail. Resistance of this load requires an evaluation assuming the entire handrail is loaded with 50 pounds (222 N) on every foot of length of the rail.

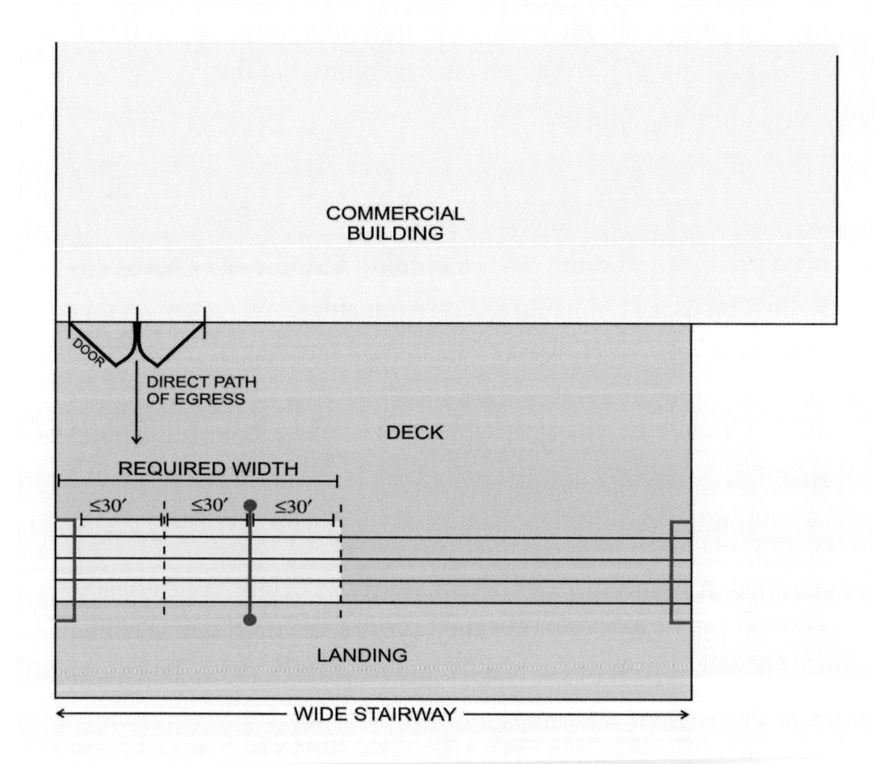

Example 9-3: This illustration depicts a stairway that is wider than the minimum required. The minimum exit width, depicted by the shaded portion of the stairs, is in the most direct path of egress travel, and must be provided a handrail within 30 inches (762 mm) horizontally from any point in the required width. A handrail must still be provided on both sides of the stairway, thus the single handrail to the far right side of the illustration.

Example 9-4: Handrail extensions at the top of a stairway are required to extend horizontal for a distance of 12 inches (305 mm). At the bottom of a stairway, the handrail must continue sloped for a horizontal distance equal to one tread depth. The horizontal portion beyond the slope at the bottom of the handrail in the photo is no longer required by the IBC.

Part Eight: Case Studies

Stand-Alone Deck at Park:

This free-standing deck at a park location is provided accessibility by the installation of a ramp. The foundation was required to have frost protection by a foundation system that extends below the frost depth. The handrails at the sides of the stairway do not provide extensions at the top and the bottom, as is required. A compliant center handrail was not required, but was installed to provide the proper handrail extensions, where not present at the sides. The guards were constructed 42 inches (1067 mm) above the deck surface. The benches secured in the middle of the deck surface had to be installed so that there was at least a 36-inch (914 mm) clear path around them.

Deck at Restaurant:

The deck added to the back of this restaurant is considered an "A" occupancy and requires a structural evaluation for a 100 pound per square foot (4.79 kPa) live load. The building is Type V construction, thus allowing the use of combustible materials in the deck construction. If the door leading to this deck is a required part of the means of egress system, the exiting from the deck must be based on the occupancy load of the building portion exiting to the deck and the deck itself. Guards are constructed at 42 inches (1067 mm) above the deck surface, and illumination of the area must be provided all the way to the base of the stair at the pavement level. The deck had to be constructed low enough that the overhead service drop attaching to the building was at least 12 feet (3658 mm) above the deck surface.

If the calculated occupant load, based on the number intended to exit the building across the deck plus the occupant load of the deck, exceeded 49 people, the gate at the top of the deck stairs would be required to swing in the direction of exiting (egress) travel, and a second exit from the deck would be required. However, the gate would not be allowed to swing over the stairway and a design modification would be required. Handrails must be provided on both sides of the stairs and must include extensions at the top and bottom. If the deck was not provided an exit directly to grade (like the stairs in this photo), and the deck occupants were required to exit through the building, the building occupant load would be increased. This occupant load increase could require modification to the existing exits from the building.

Deck at Apartment Complex:

The decks serving this apartment complex must be constructed under the provisions of the IBC, as this structure was originally built to meet the requirements of the IBC.

Appendix

The International Code Council

The International Residential Code (IRC)

This *International Code* provides comprehensive construction regulations for one- and two-family homes and townhomes, to include plumbing, mechanical, electrical, fuel-gas, building and energy conservation codes.

The International Building Code (IBC)

This *International Code* provides comprehensive building code regulations for construction of buildings of all types and uses.

International Code Council Evaluation Services (ICC-ES)

http://www.icc-es.org

This nonprofit, public-benefit corporation provides technical evaluations of building products, components, methods and materials to determine their capabilities in providing an equivalent performance to the criteria required in the International Family of Codes. Reports generated from the evaluations are provided free to the construction industry for use as evidence in a submittal to a building official for "approval."

Deck Expo

http://www.deckexpo.com

This trade show, produced by Hanley Wood Exhibitions, is exclusively dedicated to the residential deck, dock and railing industry. Product vendors and demonstrations, industry-related education and live construction demonstrations are provided in an atmosphere filled with deck-related professionals.

American Wood Council (AWC) and the American Forest and Paper Association (AF & PA)

Design for Code Acceptance Number 6

http://www.awc.org/Publications/download.html

This document provides one means to satisfy the performance requirements of the IRC as related to deck construction, but does not contain the minimum prescriptive requirements. Utilizing designs from this document would need to be approved as an "alternative."

Free eCourses

http://www.awc.org/HelpOutreach/eCourses/index.html

Many free online courses related to wood-frame construction, including a "coming soon" course for residential deck structural components.

Wood Frame Construction Manual

This book is a direct reference from the *International Residential Code* and prescriptive designs contained within do not require approval as an "alternative."

Span Tables for Joists and Rafters, 2005 edition

http://www.awc.org/pdf/STJR_2005.pdf

This free PDF for joist and rafter spans is a direct reference from the IRC and can be used without specific approval as an "alternative." The spans in this document have accounted for a wet-use environment and are acceptable for use in sizing deck joists receiving only a uniformly distributed live load.

Forest Products Society

Wood Decks

This book contains tables for sizing various structural members of a residential deck, including a table for sizing posts based on height and loading. Information in this book can be submitted for approval as an "alternative."

Southern Pine Council

http://www.southernpine.com/

This organization provides a wide variety of information about Southern Pine and the use of it in residential deck construction, including span tables which can be submitted for approval as an "alternative."

California Redwood Association (CRA)

http://www.calredwood.org/

This organization provides a wide variety of information about Redwood and its use in residential deck construction.

Western Red Cedar Lumber Association (WRCLA)

http://www.wrcla.org/

This organization provides a wide variety of information about Western Red Cedar and its use in residential deck construction.

Western Wood Products Association (WWPA)

http://www2.wwpa.org/

This organization provides a wide variety of information about softwoods, including technical publications and span tables which can be submitted for approval as an "alternative."

North American Deck and Railing Association (NADRA)

http://www.nadra.org/

This is the premier association for professionals related to the deck and railing industry, and provides various forms of assistance to such professionals.

Stair Manufacturers Association

http://www.stairways.org

This association provides an assortment of information related to the comprehension of the International Family of Codes in regard to stairway related provisions. The "Visual Interpretations of the 2006 *International Residential Code"* is a 16-page document available as a free download which includes photos and graphics intended to aid in the correct installation of stairways and handrails.

Professional Deck Builder Magazine

http://www.deckmagazine.com

This magazine, published by Hanley Wood, is distributed to deck-industry related professionals at no cost. Contained within are articles, advertisements and technical information related to the decking industry. Topics are very diverse and range from business and marketing advice to building code and carpentry techniques.

Structural Building Components Association, Wood Truss Council of America

Attachment of Residential Deck Ledger to Metal Plate Connected Wood Truss Floor System (technical note)

http://www.sbcindustry.com/images/technotes/T-DeckLedger_07.pdf

This technical report provides details for attaching a residential deck ledger to open-web wood floor trusses. Information contained within may be submitted for approval as an "alternative."

Deck Evaluation Checklist

The following is based on the Deck Evaluation Checklist published by the North American Deck and Railing Association (NADRA). The checklist intends to assist inspectors and builders in evaluating existing decks. The information reflects deck industry best practices and does not necessarily represent code requirements. The local building department should be consulted to verify code requirements.

www.NADRA.ORG

DECK EVALUATION CHECKLIST PAGE 1 OF 4

Date: ____________
Builder: ____________ Project Name: ____________
Lot Number(s) or Address(s): ____________
Superintendent: ____________ Reported By: ____________
Certified Deck Builder: Y / N

I. Ledger Connection

A. Ledger attached to an acceptable wood rim joist? ☐ Yes ☐ No ☐ Not Applicable ☐ Unable to Determine

Yes: Acceptable Rim Joist - Minimum 1" OSB Rim Joist
No: Not Acceptable Rim Joist-Siding, Stucco, I-Joists, Non-structural material (foam)
Not Applicable: Free Standing Deck or Attached to Concrete or Block

1. Fastener Type? Circle One: Lag Screws / Machine Bolts / Other: ____________
   Diameter: ____________ Length: ____________
2. Fastener Spacing: ____________ inches Staggered: ☐ Yes ☐ No
3. Any visible signs of red rust/corrosion: ☐ Yes ☐ No
   If yes, explain ____________

B. Ledger attached to concrete or CMU? ☐ YES ☐ NO ☐ NOT APPLICABLE

No: Not Acceptable or Not Allowed - Ledger Attached to Brick or Masonry Veneer
Not Applicable: Free Standing Deck or Attached to a Wood Rim Joist

1. Fastener Type? Circle One: Lag Screws / Machine Bolts / Other: ____________
   Diameter: ____________ Length: ____________
2. Any visible signs of red rust/corrosion: ☐ Yes ☐ No
   If yes, explain ____________

Notes/Comments

II. POSTS/FOOTINGS

A. Foundation type? Circle One: Footing / Pier / Other: ____________

1. Size: ____________ Depth/Thickness: ____________
2. 12" below undisturbed ground? ☐ Yes ☐ No
3. Footing meets frost protection guidelines, if required? ☐ Yes ☐ No

B. Post size? Circle One: 4x4 6x6 8x8 Other: ____________

C. Any visible signs of rot or cracks? ☐ Yes ☐ No

D. Any visible signs of red rust/corrosion? ☐ Yes ☐ No

E. Post-to-Concrete connection? Circle One: Cast-in-Place / Post-Installed

1. Cast-in-Place: Model#: ____________ Fasteners: ____________
2. Post-Installed: Model#: ____________ Fasteners: ____________ Anchor: ____________

NOTES/COMMENTS

www.NADRA.ORG

DECK EVALUATION CHECKLIST PAGE 2 OF 4

III. Post-to-Beam Connection

A. **Post size?** Circle One: 4x4 6x6 8x8 Other:________________

B. **Beam size?** Circle One: 2-2x 3-2x 4x 6x Other:________________

C. **Any visible signs of rot or cracks, especially if post is notched?** ☐ Yes ☐ No

D. **Post-to-Beam connection?** ☐ Yes ☐ No Model#: ____________

1. Has the Post-to-Beam connection been bent or modified? ☐ Yes ☐ No
2. Are all the connector holes filled with the proper fastener? ☐ Yes ☐ No
3. Are the girders alongside the post? ☐ Yes ☐ No
   If yes, is the girder attached with metal connector providing bearing? ☐ Yes ☐ No
4. Are multiple members (built-up lumber) fastened together to act as a single unit?
   ☐ Yes ☐ No How?______________ Fastener type? Circle: Nails / Bolts / Screws
5. Any visible signs of red rust/corrosion: ☐ Yes ☐ No
   If yes, explain ____

Notes/Comments

IV. Joists and Joist Connections

A. **Joist:** Size:____________ Spacing:______________ Span:____________

B. **Joist hangers:** ______________ ______________

C. **Any visible signs of red rust or corrosion on the connectors or nails?** ☐ Yes ☐ No

D. **Has the hanger been bent or modified?** ☐ Yes ☐ No

E. **Does the hanger have "double-shear" nailing, see figure "A" ?** ☐ Yes ☐ No

F. **If the hanger has "double-shear" nailing, was the correct (full length) nail used for the joist into header nail?**
☐ Yes ☐ No (0.148 x 3" or 0.162 x 3½")
Hint: Do NOT use short 1½" nails for double-shear nailing as shown in figure "A".

G. **Are the correct nails installed in the hangers?** ☐ Yes ☐ No
Circle One: 0.148 x 3" HDG or SS (10d common),
0.162 x 3½" HDG or SS (16d common),
0.148 x 1½" HDG or SS

FIGURE "A"

H. **Is there a connection at the point where the joist bears on the top of the beam?** ☐ Yes ☐ No **Type:** ________________

Notes/Comments

www.NADRA.ORG

DECK EVALUATION CHECKLIST

PAGE 3 OF 4

V. STAIRS

A. Stair Rise: ______________

1. Is the gap between treads less than 4"? ☐ Yes ☐ No

B. Stair Run: ______________

C. Solid Stringer: ☐ Yes ☐ No

1. If, solid stringer; what connection supports the stairs? ______________
   Blocking/Fasteners______________ Hardware______________

D. Notched Stringer: ☐ Yes ☐ No

1. Any visible signs of rot or cracks? ☐ Yes ☐ No
2. What is the span of the stringer? ____ft. ____in.
3. Does the triangular opening formed by the riser, tread & bottom of the guard (if present) create a gap > 6"? ☐ Yes ☐ No

E. Stringer:

1. What is supporting the stringer? Hardware______________ Other______________
2. What is the spacing of the stringer? ______________ inches

Notes/Comments

VI. DECK BOARDS

A. What type of decking boards? Circle One: Wood / Composite

1. Any visible signs of rot or cracks? ☐ Yes ☐ No
2. Any nails or screws exposed? ☐ Yes ☐ No
3. Fastener Type? Circle One: Nail / Screw / Hidden

B. If Composite Decking:

1. Gap per manufacturer's guideline? ☐ Yes ☐ No ☐ Don't Know Gap Spacing ______
2. Spacing of Joists per manufacturer's guideline? ☐ Yes ☐ No ☐ Don't Know
   Joist Spacing ______________

C. If Hidden Fastener System; what lateral support has been provided?
Circle One: Cross Bracing / Angled Braces / Other:______________

Notes/Comments

VII. HANDRAIL ASSEMBLIES AND GUARDS

A. Guardrail Height? Circle One: 36" / 42" / Other:______________

B. Is there a Guardrail Post? ☐ Yes ☐ No Guardrail Post Spacing?______________

C. Is there a shear connection between the post and the frame? ☐ Yes ☐ No

1. If yes, what type? Hardware ______________
2. If no, Circle One: Bolts, only / Lag Screws, only / Other ______________

D. Is the guardrail post notched? ☐ Yes ☐ No

E. Any visible signs of rot or cracks? ☐ Yes ☐ No

F. Any signs of corrosion or rust in the hardware? ☐ Yes ☐ No

G. Is the opening between the balusters on the deck less than 4"? ______________

H. Is the opening between the balusters on the stairs less than 4 3/8"? ______________

I. Is the handrail graspable? ☐ Yes ☐ No

J. Does the handrail return to a post or safety terminal? ☐ Yes ☐ No

Notes/Comments

www.NADRA.ORG

DECK EVALUATION CHECKLIST

PAGE 4 OF 4

VIII. MISCELLANEOUS

A. Any signs of corrosion/red rust, not previously mentioned? ______

B. All fasteners properly seated (flush with the connection) ? ______

C. All connector holes properly filled (fill all round & obround holes)? ______

D. All bolt holes drilled 1/32" to no greater than 1/16" larger than the bolt diameter? ______

E. All bolts have washers on the wood side of the connection? ______

F. Proper Finish for Hardware?

1. Connectors-circle

   G185 / Stainless Steel (316)

2. Fasteners-circle

   HDG mtg. ASTM A 153 / 304 SS / Other:______

3. Anchors-circle

   Hint: Zinc plated or uncoated may not be acceptable, 304 SS may not be acceptable for applications which are considered at risk for chloride-related corrosion.

   HDG / Mechanically Galvanized (MG) / 304 SS / SS

Notes/Comments

Additional Comments:

North American Deck and Railing Association

PO Box 829 • Quakertown, PA 18951 • 1.888.623.7248 • info@NADRA.org

Deck Evaluation Form: http://www.nadra.org/education.html

Find A Builder: http://www.nadra.org/find_deck_builders.html

Deck Safety: http://www.nadra.org/consumers/deck_safety_month.html

Deck For A Soldier: http://www.nadra.org/consumers/deck_safety_month.html

Photo Gallery: http://www.nadra.org/consumers/deck_4_soldier.html

ALL INFORMATION PROVIDED SHOULD BE EVALUATED BY A QUALIFIED PROFESSIONAL AND APPROVED BY THE BUILDING DEPARTMENT. EVALUATION OF THE DECK USING THIS INFORMATION DOES NOT COMPLETELY CONSTITUTE A CODE COMPLIANT DECK. IT IS INTENDED TO ASSIST BUILDERS AND INSPECTORS IN THE DECK EVALUATION PROCESS.